D0271883

MATHEMATICAL MODELLING

Mathematical Modelling

Edited by

J.G. Andrews
CEGB, Marchwood Engineering Laboratories and Department of Mathematics, University of Southampton

and

R.R. McLone
Department of Mathematics, University of Southampton

BUTTERWORTHS
LONDON — BOSTON
Sydney — Wellington — Durban — Toronto

THE BUTTERWORTH GROUP

UK

Butterworth & Co (Publishers) Ltd
London: 88 Kingsway, WC2B 6AB

AUSTRALIA

Butterworths Pty Ltd
Sydney: 586 Pacific Highway, Chatswood, NSW 2067
Also at Melbourne, Brisbane, Adelaide and Perth

SOUTH AFRICA

Butterworth & Co (South Africa) (Pty) Ltd
Durban: 152-154 Gale Street

NEW ZEALAND

Butterworths of New Zealand Ltd
Wellington: 26-28 Waring Taylor Street, 1

CANADA

Butterworth & Co (Canada) Ltd
Toronto: 2265 Midland Avenue, Scarborough, Ontario, M1P 4S1
Scarborough, Ontario, M1P 4S1

USA

Butterworths (Publishers) Inc
Boston: 19 Cummings Park, Woburn, Mass. 01801

All rights reserved. No part of this publication may be reproduced or transmitted
in any form or by any means, including photocopying and recording, without the
written permission of the copyright holder, application for which should be
addressed to the publisher. Such written permission must also be obtained before
any part of this publication is stored in a retrieval system of any nature.

This book is sold subject to the Standard Conditions of Sale of Net Books and
may not be re-sold in the UK below the net price given by the Publishers in
their current price list.

First published 1976

ISBN 0 408 10601 8

Chapters 1-12 and 14-17
© Butterworth & Co (Publishers) Ltd 1976
Chapter 13
© R.E. Beard 1976

LIBRARY OF CONGRESS CATALOGING IN PUBLICATION DATA

Main entry under title:

Mathematical modelling.

   Bibliography: p.
   Includes index.
   1. Problem solving. 2. Mathematical models.
I. Andrews, J.G. II. McLone, Ronald Redman.
QA63.M36   511'.8   76-23335
ISBN 0-408-10601-8

Printed in England by Chapel River Press, Andover, Hants.

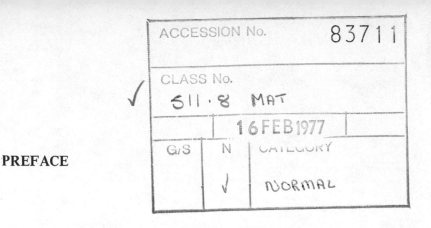

ACCESSION No.   83711

CLASS No.

✓   511·8   MAT

16 FEB 1977

G/S | N | CATEGORY

✓ | NORMAL

PREFACE

The traditional undergraduate degree course in mathematics provides most students with a formal approach to their subject. It is taught in a developed way emphasising basic principles and a need for rigour. However, these features form only part of the armoury of a practising mathematician. A common complaint amongst employers is that prospective mathematical recruits are often too inhibited to tackle raw problems as they arise. This is not necessarily due to any lack of ability, but is perhaps a reflection on the formal nature of mathematics degree courses compared with other disciplines.

One possible answer to this problem is to introduce into the traditional degree a course in which the major aim is to consider the *formulation* of mathematical problems from real practical situations. A broad title for this approach is 'mathematical modelling'. No universally accepted definition of this phrase exists at present, but differing types of mathematical 'applicators' are usually clear what it means within their own area. The aim of this book is to bring together the many different aspects of mathematical modelling through case studies from a wide range of applications. It is not essential, therefore, to follow the book chapter by chapter; though the reader may be guided by the outlines described in chapter one.

In practice, such a course can only have value for individual students provided they carry out some development of models for themselves. Both 'closed' and 'open-ended' problems are set at the end of each chapter. The 'closed' questions are designed primarily to test comprehension of the case study and the 'open-ended' questions provide an opportunity for the student to develop his own ideas.

A modelling course could in principle be given at any stage in the undergraduate's career and some of the examples in this book require little more than school mathematics. On the other hand, most chapters do require a deeper mathematical appreciation and hence the book is more suitable as a basis for second or third year undergraduate courses in mathematical modelling.

J.G.A.
R.R.McL.

v

GLOSSARY OF MAIN SYMBOLS

Chapter 2 (Steering Problems)

$\mathbf{x}$ = $(x_1, x_2, ..., x_n)$ vector equation of path
$\boldsymbol{a}$ = $(a_1, a_2, ..., a_m)$ vector rotation of steering-wheel

Chapter 3 (Why Build Three-Stage Rockets?)

m	mass
r	distance from centre of earth
R	radius of earth
g	acceleration due to gravity
v	velocity of rocket relative to earth
u	velocity of exhaust gases relative to rocket
T	thrust on rocket
m_0	initial mass of rocket
m_P	mass of payload
m_F	initial mass of fuel
m_S	mass of structure
λ	ratio of mass of structure to combined mass of fuel plus structure

Chapter 4 (Liquid Flowing from a Container)

ρ	density of milk
p	pressure
u	speed of fluid
V	volume of fluid flowing out of container per unit time
A	cross-sectional area
C = A_B/A_0	contraction ratio, i.e. ratio of cross-sectional area of fully developed jet, A_B, to that of the hole, A_0.
a	angle made by side of container to axis of symmetry
y	height of free surface above hole
t	time
V_0	initial volume of fluid in container

h	initial value of y
λ	decay constant for transient effects
θ	angle between side AB and the vertical
a	length of side AB
b	length of side BC
T	total emptying time

Chapter 5 (Molecular Models)

T	topological matrix
B	adjacency matrix
N	total number of C–C links
Tr	trace operation
r	number of rings
n	total number of carbon atoms
R	resonance energy
S	number of carbon atoms bonded to one other carbon
P	set of orthonormal vectors combined to form a matrix
d	internuclear distance
p	bond order
V	location of unpaired electron

Chapter 6 (Drilling Holes With a Laser)

W	power
A	area of heated surface
$s(t)$	depth of hole as function of time
v	speed of boundary in evaporation-controlled limit (no heat conduction)
h	heat required to evaporate unit mass of material
ρ	density of material
T	temperature
z	co-ordinate measured from upper surface of material
D	diffusivity of material
K	thermal conductivity
c	specific heat
T_v	boiling point
L_v	latent heat of evaporation (per unit mass)
l	characteristic length for temperature decay
$\epsilon = cT_v/L_v$	
$\theta = T/T_v$	normalised temperature
$\zeta = z/l$	normalised spatial coordinate
$\tau = vt/l$	normalised time
$\xi = s/l$	normalised position of moving boundary

Chapter 7 (Stress Analysis – Structures and the Birth of the Finite-Element Method)

f	force
u	displacement
K_{ij}	coefficient of elasticity
σ	stress
b	body force
L	differential operator for forces acting on material
ϵ	strain vector

Chapter 8 (Population Models)

N_t	size of population at time t
N_0	initial size of population
b	birth rate
d	death rate
$r = b - d$	'net birth rate'
K	equilibrium size of population
R	rate of increase of population between generations
l_x	probability of individual surviving to age x

Chapter 9 (A Differential Model of Diabetes Mellitus)

x	blood sugar level
y	blood insulin level
z	food input
w	insulin input
t	time
$H(\xi)$	unit step function, $H = 0$ for $\xi < 0$, $H = 1$ for $\xi \geqslant 0$
Q	quantity of food
K	delay parameter
t_0	time of meal

Chapter 10 (Stochastic Models for Road Traffic Situations)

t	time
T	given time gap
λ	mean number of vehicles per unit time
$P(k)$	Poisson distribution function
a	minimum time gap
$f(t)$	probability density function
$F(t)$	probability distribution function
$f^*(s)$	Laplace transform of $f(t)$

$H(u)$ probability distribution function for sum, u, of two gaps
$w(t)$ probability density function for delay of vehicles in minor road

Chapter 11 (A Business Planning Model)

S_t number of installations in year t
u_t charge per installation in year t
T_{it} number of customers of rental class i in year t
r_{it} rent per customer of rental class i in year t
D_{jt} number of calls made in charge category j in year t
x_{jt} charge per call in category j in year t
J_t total income in year t
V_t income from other services
E manpower expenditure
P_k manpower productivity in service category k
W_k wage rate in service category k
G gross asset
D depreciation
r depreciation rate
N net value of assets at end of year
C net capital expenditure in year
I_t interest paid on loans in year t
B_t money borrowed in year t

R_t $\displaystyle\prod_{v=1}^{t} r_v$

Chapter 12 (The Control of the Grade Structure in a University)

$n_i(T)$ number of people in grade i at time T ($i = 1,2,...,k$)
$n_{ij}(T)$ number of people moving from grade i to j between times T and $T + 1$
$n_{0i}(T+1)$ new entrants into grade i between times T and $T + 1$
$n_{jj}(T)$ number of survivors in grade j
p_{ij} probability of moving from grade i to grade j between years T and $T + 1$
w_i probability of leaving the structure (completely) from grade i
$R(T+1)$ total recruitment at $T + 1$
r_i constraint on grade i
$Q = P + w'r$ stochastic matrix representing all possible grade-to-grade transitions
x vector proportion of people in particular grade

Chapter 13 (A Mathematical Model for Motor Insurance)

U_0 initial fund
U_t fund after time interval t
P_t premiums received in time interval t
C_t claims incurred in time interval t
μ number of claims per unit time
n total number of expected claims
$S(x)$ probability distribution function for one claim of amount x
$F(x,n)$ probability distribution function of total claims
ϵ probability of ruin
λ security loading factor

Chapter 14 (A Military Application of Game Theory)

F_{ij} probability that a mine on ship-count i will sink a ship passed through the channel after j passes of the sweeper
p_i probability of ship-count i being used
q_j probability of making j passes
P,Q associated probability distribution functions
Y total length of channel
y length coordinate of channel
N expected number of mines in channel
$P(y)$ probability of ship reaching a distance y undamaged
R risk, i.e. probability of being sunk
M number of ships entering channel
μ_{ij} expected number of ships sunk

Chapter 15 (Network Flow Models)

$G = (N,A)$ network (or graph)
$N = \{n_1, n_2, ..., n_p\}$ set of nodes
A set of ordered pairs of arcs (n_i, n_j)
$c(n_i,n_j)$ capacity of arc (n_i, n_j)
v network flow
$\phi(n_i,n_j)$ arc flow in arc (n_i, n_j)
$\ell(n_i,n_j)$ cost of flow in arc (n_i, n_j)
μ path

Chapter 16 (Urban Structure)

A set of activities
B set of buildings
P set of people
Λ incidence matrix relating sets of people with sets of activities

X, Y arbitrary sets

K simplicial complex

σ simplex

$N = \dim K$ dimension of K

γ_q relation on the simplices of K at level q

Q_q cardinality of K/γ_q, equal to the number of distinct q-connected components in K

π pattern on a simplicial complex K, i.e. the mapping $\pi : K \to J$

f *face* operator which replaces a simplex σ_p by all its $(p - 1)$ faces

Chapter 17 (Structural Stability of Mathematical Models: The Role of the Catastrophe Method)

$x(t) = (x_1(t), x_2(t), ..., x_n(t))$ vector state of a system at time t

V potential function of system

S rate at which substance enters a cell

m mass of simple pendulum

ℓ length of simple pendulum

θ angular displacement of pendulum

$c = (c_1, c_2, ..., c_k)$ set of parameters

V_c potential function for a particular set of parameters, c

K catastrophe set, i.e. set of parameters c such that V_c has some coalescent critical points

CONTENTS

1   **Mathematical Modelling – The Art of Applying Mathematics**   1
R.R. McLone, Department of Mathematics, University of Southampton
1.1   Why 'Modelling'?   1
1.2   General features of modelling   2
1.3   The models   6
1.4   References   10

2   **Steering Problems**   12
J.W. Craggs, Department of Mathematics, University of Southampton
2.1   The problem   12
2.2   Restricted domains   14
2.3   Motion with a preferred direction   15
2.4   Motion with restricted derivatives   17
2.5   Manoeuvring a car   20
2.6   Other applications   22
2.7   Reference   23
2.8   Problems for further study   23

3   **Why Build Three-Stage Rockets?**   26
B. Noble, University of Wisconsin and University of Oxford
3.1   Identification of the problem   26
3.2   The motion of the satellite   27
3.3   The force exerted by a rocket engine   29
3.4   The mass of a rocket-satellite system   31
3.5   Practical realisation of the ideal performance   33
3.6   Reference   37
3.7   Problems for further study   37

4   **Liquid Flowing from a Container**   39
N. Curle, Department of Applied Mathematics, University of St. Andrews
4.1   Introduction   39
4.2   The Helmholtz–Kirchoff free streamline theory   39
4.3   Emptying time for arbitrary fixed container   42

4.4	Application to milk sachet	44
4.5	The effects of the starting period	46
4.6	The effects of optimum tilting of the container	48
4.7	Time taken to tilt container	52
4.8	References	54
4.9	Problems for further study	54

5 Molecular Models   **56**
G.G. Hall, Department of Mathematics, University of Nottingham

5.1	Introduction	56
5.2	Planar hydrocarbon molecules	56
5.3	Graph and matrix models	57
5.4	Free radicals	59
5.5	Structures	59
5.6	Enumeration of structures	62
5.7	Resonance energy	63
5.8	Bond orders	65
5.9	Internuclear distances	68
5.10	Zero eigenvalues	68
5.11	Reference	69
5.12	Problems for further study	69

6 Drilling Holes With a Laser   **71**
J.G. Andrews and D.R. Atthey, CEGB, Marchwood Engineering Laboratories, Southampton

6.1	Introduction	71
6.2	Basic physical model	73
6.3	More accurate mathematical model: allowance for heat conduction	74
6.4	Simple perturbation solutions	76
6.5	Discussion	79
6.6	References	80
6.7	Problems for further study	81

7 Stress Analysis – Structures and the Birth of the Finite-Element Method   **83**
O.C. Zienkiewicz, Department of Civil Engineering, University of Wales, Swansea

7.1	Introduction	83
7.2	The discrete problem	86
7.3	The plane stress analysis problem and the first finite element	87
7.4	Some obvious questions and the generalisations	91
7.5	Concluding remarks	95
7.6	References	95
7.7	Problems for further study	96

8   **Population Models**   98
    G. Murdie, Department of Zoology and Applied
    Entomology, Imperial College, London
    8.1   Introduction   98
    8.2   Single-species populations   99
    8.3   Two species models   104
          8.3.1   Competing species   104
          8.3.2   Host-parasite systems   105
    8.4   References   113
    8.5   Problems for further study   114

9   **A Differential Model of Diabetes Mellitus**   116
    M.J. Davies, Applied Mathematics Department,
    University College of Wales, Aberystwyth
    9.1   Introduction   116
    9.2   Variable identification   116
    9.3   State relations   117
    9.4   The source terms   119
    9.5   The analysis   120
    9.6   Discussion   120
    9.7   References   125
    9.8   Problems for further study   125

10 **Stochastic Models for Road Traffic Situations**   127
    Winifred D. Ashton, Department of Mathematics,
    University of Surrey
    10.1 Models for road traffic   127
    10.2 A simple model for pedestrian delay   128
    10.3 A simple model for a priority intersection   132
    10.4 More complicated models for intersections   136
    10.5 Headway distributions   137
    10.6 Gap-acceptance models   140
    10.7 Validation of the model   140
    10.8 References   141
    10.9 Problems for further study   141

11 **A Business Planning Model**   143
    T. Lomas, Management Services Department, Post Office
    Telecommunications Headquarters, London
    11.1 Introduction   143
    11.2 Major areas of the model   144
    11.3 Manpower sub-model   146
    11.4 Depreciation sub-model   146
    11.5 Financing sub-model   147
    11.6 Use of the model in tariff decisions   151
    11.7 Example of tariff calculations   156
    11.8 References   159
    11.9 Problems for further study   159

12 The Control of the Grade Structure in a University   161
    D.J. Bartholomew, Department of Statistics, London
    School of Economics and Political Sciences
    12.1 The problem   161
    12.2 Stocks and flows   162
    12.3 Assumptions about flows   162
    12.4 The basic prediction equation   164
    12.5 Prediction   165
    12.6 Control: maintainability   167
    12.7 Control: attainability   170
    12.8 Concluding remarks   171
    12.9 Bibliography   172
    12.10 Problems for further study   172

13 A Mathematical Model for Motor Insurance   174
    R.E. Beard, Department of Trade and Industry and
    Department of Mathematics, University of Essex
    13.1 Introduction   174
    13.2 Criterion for non-ruin   175
    13.3 Compound Poisson process   175
    13.4 Limitations of the model   176
    13.5 Reinsurance   177
    13.6 A numerical example   178
    13.7 References   181
    13.8 Problems for further study   181

14 A Military Application of Game Theory   183
    W. Hill, AUWE, Portland, Dorset
    14.1 Introduction   183
    14.2 Two-person games   184
    14.3 Measures of effectiveness   184
    14.4 Mine warfare   185
    14.5 The basic game   185
    14.6 First extension: many mines   188
    14.7 Second extension: mixtures of strategies   190
    14.8 Third extension: many ships   193
    14.9 Final extension: low loss criterion   194
    14.10 Summary   196
    14.11 References   196
    14.12 Problems for further study   196

15 Network Flow Models   199
    B.A. Carré, Centre d'Information, Université Paul Sabatier,
    Toulouse, France and Department of Electronics,
    University of Southampton
    15.1 Introduction   199
    15.2 Networks   200
    15.3 Flows in networks   201
    15.4 A method of finding maximal flows   202

15.5 Minimal-cost flows — 206
15.6 Some network flow models — 208
15.7 Extensions — 212
15.8 Alternative modelling methods — 213
15.9 References — 214
15.10 Problems for further study — 215

16 Urban Structure — 217
R.H. Atkin, Department of Mathematics, University of Essex
16.1 Introduction — 217
16.2 Mathematical relations in an urban community — 218
16.3 The structure of a relation — 220
16.4 Some consequences — 223
16.5 References — 230
16.6 Problems for further study — 230

17 Structural Stability of Mathematical Models: The Role of the Catastrophe Method — 231
D.R.J. Chillingworth, Department of Mathematics, University of Southampton
17.1 Introduction — 231
17.2 Systems governed by a potential function — 233
17.3 Critical points and structural stability — 236
17.4 Equilibria for systems controlled by k parameters — 238
17.5 The delay situation and the Maxwell situation — 242
17.6 The main theorem — 244
17.7 Thom's list of seven catastrophe models — 245
17.8 Use of the catastrophe models in application — 253
17.9 Conclusion — 254
17.10 References — 255
17.11 Problems for further study — 256

Index — 259

1
MATHEMATICAL MODELLING –
THE ART OF APPLYING MATHEMATICS

R.R. McLone
Department of Mathematics, University of Southampton

1.1. Why 'Modelling'?

Most mathematics students are exposed to the development of highly
formalised theories in particular fields, and to the accumulation of a
number of mathematical techniques. Both these aspects of undergraduate mathematics are important and form part of the stock in
trade of an applied mathematician*. The study, for example, of
Newtonian mechanics allows one to see an accepted model elegantly
displayed. However, by such study the student does not see how
Newton had to struggle to devise and produce what is now an established theory. Equally it does not show how this model is relevant
to the practical problems facing engineers and physicists· every day in
the industrial situation. Thus, one obtains the impression that the
application of mathematics involves simply the 'looking up' of appropriate formulae, the substitution of some numbers and a magic stir
from which comes the 'answer'. In this way an important ingredient
is left out without which the 'application' becomes simply another
piece of mathematical technique.

What is the missing ingredient? It is the representation of our
so-called 'real world' in mathematical terms so that we may gain a
more precise understanding of its significant properties, and which may
hopefully allow some form of prediction of future events. This has

*In this book the term 'applied mathematician' refers to anyone who applies
any branch of mathematics to describe the real world, whether that be statistics,
operational research, numerical analysis, computation, or the more traditional
areas of pure and applied mathematics.

1

been described in the term 'mathematical modelling'. In this book we shall attempt to give the undergraduate an appreciation of this aspect of mathematical education.

In a recent survey (McLone, 1973) of both employers of mathematicians and those who had graduated with mathematics degrees from UK universities, it was commonly felt that modelling was overlooked in many traditional university courses. The areas of problem formulation, development of new ideas and extension of these or existing theories to new areas of application, criticism of developed pieces of work, and communication and interpretation of conclusions to others (especially non-mathematicians) are generally only briefly covered in the undergraduate mathematics curriculum, whilst these abilities (among others) are regarded as central to the student's education. We are concerned to illustrate how these skills and abilities are involved in the carrying out of mathematical modelling by providing a number of situations from many different fields which the various authors have attempted to model mathematically.

1.2.   General features of modelling

Hall (1963) has said 'The goal of applied mathematics is to understand reality mathematically'. On the other hand, the practising engineer may well be more concerned to know whether his bridge will withstand the likely load to be placed on it, and the hospital administrator to obtain a way of reducing the waiting time for patients at the local out-patient clinic – that is, to obtain specific answers to specific questions – than to strive towards such high ideals.

Where does the applied mathematician start? In practice the start is often in some empirical situation which presents a 'problem' for which an 'answer' is sought. However, the use of such words as 'problem' and 'answer' can be very misleading. Firstly one has to identify what the 'problem' really is; this recognises that real situations rarely appear well-defined and are often embedded in an environment which makes a clear statement of the situation hard to obtain. The identification of a 'problem' amenable to mathematical treatment is often long and involves many skills which are not related to mathematics; for example, talking to non-mathematical colleagues in the problem area and reading any relevant literature are important features of this part of the modelling exercise.

Often (but not always) at the same time that this identification stage is being carried out, a process of sorting out the essential or significant features takes place (*see Figure 1.1*). In physical situations in particular, this *simplification* or *idealisation* is a crucial stage since the general problem is usually exceedingly complex, involving many processes. Some features will appear significant, many irrelevant. For example, consider the motion of a pendulum formed by attaching a heavy weight to the end of a piece of string. When analysing this

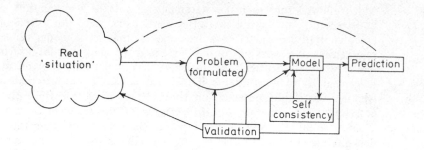

Figure 1.1

'situation' it may be regarded as significant that the swing repeats
itself regularly, but irrelevant that the string is white or the weight
painted black. Once the significant features have been identified, the
next stage is to translate these into mathematical entities, and to
postulate relations between these entities. This is generally the most
difficult stage and one in which it is impossible to give formal instruc-
.tion! The modelling described in this book illustrates the variety of
approaches open to the mathematician.

Once a model is constructed it needs to be validated. Indeed,
some form of validation is usually carried out throughout the formulation; the equations or other mathematical relations set up in the model
are continually checked with the initial situation. For example, in the
case of the pendulum just described, the mathematical equation for the
motion of the pendulum may be checked by reference to the physical
dimensions of the quantities thus represented.

Validation takes a number of forms. In the first place, the mathematics itself which constitutes the model must be self-consistent and
obey all the usual laws of mathematical logic. In the second place,
a model's validity rests in its ability to represent the situation initially
described; however, whether a model passes this test is to a large
extent subjective. A model may have to represent reality, but it is
not *itself* reality. Our pendulum is real enough; but it is often said
to be a simple pendulum and that is precisely the easy trap waiting
for us to fall into. For what is known to physicists and applied
mathematicians alike as a simple pendulum is but a mathematical
idealisation of the real object and no more. This becomes clear as
we see the swing of our pendulum decrease in magnitude until ultimately the pendulum appears at rest; the simple pendulum model does
not predict such an occurrence. Is then the simple pendulum model
invalid? Not necessarily, since it may take an hour or more for the
swing to die away and maybe we are only interested in events during
the first five minutes.

'Validity' also appears in other forms. For example, in describing
the operation of an appointments system at an out-patient clinic,

Shahani (1974) adopted a standard approach involving queuing theory which allowed him to make statements about the waiting time of both patients and consultant. His conclusions were not based on the acceptance of his model as a *true* model, that is, a model accurately representing the working of the system at all stages; indeed such a model would be difficult to justify. They were simply based on a judgement as to whether the model was *adequate* or not; that is, whether the results obtained on the basis of it were sufficiently representative of the situation *for the purpose of the problem in hand*. Thus a 'solution', even of the same problem at a different time, depends on criteria assessed by the modeller as much as on identification of physical (or economic, or any other) features of the initial situation.

Much time can. be wasted in refining a solution to a model to an extent not justified by the formulation of the problem itself. An aspect of this concerns the accuracy of empirical data. Thus, if the original data provided are liable to an error of, say, 5 per cent, it is clearly meaningless to present 'solutions' which are expected to be accurate to within 1 per cent. As a corollary, it is worth emphasising that an answer which, although based on sophisticated mathematical treatment, is impossible to realise in practice, is irrelevant to the problem at hand. As one engineer once said 'Any equation more than two inches long is probably wrong'! It is also the case that an approximate answer that can be obtained quickly may be more effective than a more accurate answer that takes a long time – which often means a straightforward numerical approximation without spending time looking for the most elegant analytical solution.

Situations are modelled for various purposes. Foremost among these is the need to predict new results or new features, which may be of the form of extensions of the existing results, or of a more radical nature. The predictions are often of conditions likely to exist at some future date. They may, on the other hand, be predictions of events for which direct experimental evidence is unobtainable; an extreme example of this lies in the many predictions made on the basis of mathematical models in the space programme. Not all situations are modelled for this purpose however; in some it is sufficient to be able to describe the working of the system by mathematical means in order to obtain a greater understanding – many of the great physical theories do this, although they make predictions as well. What is not usually involved in such mathematical descriptions is an element of control, but in the models constructed, for example, to investigate a network operation such as a routing schedule for trains or aircraft, control is often an important consideration. Indeed, many models in operational research or engineering are designed to assist management in the making of *decisions*. Other models may be constructed to facilitate a scale of measurement. Thus a linear scale for temperature is a mathematical model which is used in the development of a measuring device, namely a thermometer; many other examples can be found.

The mathematical model represents a *simplification* of the actual situation. A recognised stage of simplification is obtained when, by ignoring insignificant features, an originally complex problem is idealised to one which is mathematically tractable. In this way classical applied mathematics has given birth to frictionless pulleys, light inextensible strings, inviscid fluids and many others of the same *genre*. Such concepts have no real existence; they are abstractions, part of the modeller's idealisation which nevertheless can often usefully be regarded as a close approximation to real-life situations.

This is just one aspect of simplification; another is concerned with the relative magnitude of different quantities involved in the model. For example, the variation of some quantity x with time may be expressed in the form of an equation:

$$a \frac{\mathrm{d}^2 x}{\mathrm{d}t^2} + b \frac{\mathrm{d}x}{\mathrm{d}t} + cx = 0$$

Now one may proceed immediately to the solution of such an equation; but suppose, by observation, it is seen that $b\mathrm{d}x/\mathrm{d}t$ is much greater in magnitude than cx? Much time may then be saved by a simplification which leads to a *quicker* solution (by ignoring the cx term) and yet gives a solution which is representative of the situation. Indeed, the solution to the original equation, whilst being mathematically more exact, may lead to misleading conclusions.

This way of proceeding in the construction of mathematical models is not unique, nor should it be expected to be so. Another philosophy starts by constructing a simple model of a few readily evident features, often to obtain a 'feel' for the problem in hand, and before the problem itself is completely formulated. This simple model is then extended by adding other factors until an 'acceptable' or 'adequate' solution is found. A further alternative involves the consideration of a large number of factors *ab initio*, on a multivariate basis. This approach is often applied in operational research, and such models usually require solution by simulation methods using an electronic computer.

A major decision, often taken at the very onset of modelling concerns the nature of the mathematical variables involved. Basically variables are of one of two types. One class represents *known* features, that is quantities capable, at least in theory, of precise measurement and control; they are called *deterministic* variables. The other represents *unknown* features, that is quantities the measurement of which can never be known precisely, but which illustrate a random feature; these are called *stochastic* variables. A model which contains stochastic variables must by definition be concerned with mathematical techniques appropriate to probability and statistics; deterministic variables often, but by no means exclusively, call for the use of the calculus. Some situations do not reveal their nature at the onset; others have variables of both types in them. It is fundamental to the construction of the model that the nature of the variables be properly diagnosed.

Finally, we are led to the question of the *interpretation* of the results of the model. An applied mathematician's task is not complete when, after much mathematical manipulation and derivation, a formula or other statement is obtained. It remains to re-interpret the mathematics in terms of the problem as it originally arose — non-mathematical colleagues in management, for example, are unlikely to be impressed by answers in a language they do not understand. The implications of the solutions, both for the mathematics and in terms of the reality it is supposed to represent, should be clearly spelt out. This is a feature not always underlined in traditional undergraduate mathematics courses.

The development of a mathematical model may be regarded as analogous to that of a young child learning to speak. For in a very real sense, language is a representation of the real world, and the development of language in the child is related to his understanding of that world. The model starts simply and develops into something more sophisticated as greater understanding is achieved.

1.3. The models

The art of model building can only be acquired through first hand experience but it can be appreciated by the study of examples that illustrate the various features in one way or other. The wide range of mathematical applications can be illustrated from books on mathematical biology, economics, politics, psychology etc. (Smith, 1968; Rosen, 1967; Alker, 1965; Miller, 1964); a number of more general articles on modelling itself underline the general principles (Woods, 1969; Ford and Hall, 1970; Hall, 1972; Klamkin, 1970; Wallis, 1971).

The models described in this book fall generally into four groups. The first group, namely those by Craggs, Noble, Curle, Hall, Andrews and Atthey, and Zienkiewicz are broadly models of physical situations, although the last is largely concerned with a mathematical tool of great use to the structural engineer. The second group, Murdie and Davies, are models related to biological and medical situations respectively, although in company with the first group they are *deterministic* in nature. The third group, Ashton, Lomas, Bartholomew, Beard and Hill are models which rely on statistical and operational research methods; they are *stochastic* in nature. The final group, Carré, Atkin, and Chillingworth are models different in kind from the rest, and illustrate an essentially mathematical approach. We conclude this chapter with a description of the essential modelling aspects of each contribution in relation to the general features enunciated in the previous section.

CHAPTER 2 (CRAGGS)

This illustrates a series of situations concerning the steering of a 'body'
from one place to another. The author has chosen to model the
purely geometrical aspects; the real life situation will involve many
other features of which this is just one. It is important for the
modeller to assess under what conditions the purely geometrical model
is adequate.

CHAPTER 3 (NOBLE)

This shows how an engineering problem can be treated as a simple
application of a well-established model, i.e. Newtonian mechanics.
A complete description of rocket motion within the earth's atmosphere
would, of course, include the weight of the rocket, air resistance etc.;
Noble's model explicitly ignores these factors.

CHAPTER 4 (CURLE)

This contribution identifies in a practical situation a standard simplified
problem in fluid dynamics which is then solved. As in Chapter 3, the
complete specification of the problem would be very complex; the
essential modelling feature here resting on the observation that the
cross-sectional area of the jet is small compared with the cross-sectional
area of the fluid in the container, which allows the assumption of
negligible sideways velocity distribution. The degree of interest in such
a problem would not justify further refinements and therefore the
solution is accepted. The model is an idealisation, but a useful one
of a complicated 'real' situation.

CHAPTER 5 (HALL)

Certain physical and chemical characteristics of molecules (such as
molecular bonding) are represented by a simple topological framework
using matrices. This simplified picture allows understanding of the
chemical characteristics without detailed analysis (for example by quantum theory). The detailed analysis would of course be yet another
model; the model followed would depend on the accuracy required,
the amount of detail needed, and the time available for work on the
problem.

CHAPTER 6 (ANDREWS AND ATTHEY)

This is an example of an engineering situation involving the possibility of a large number of physical processes. A regime of interest is identified where one particular physical process (heat transfer) dominates, and this is modelled with the aid of standard established theories. The model is constantly checked for self-consistency to justify the neglect of the other possible factors. The initial identification of a regime of interest required much consultation with those in the originating engineering situation, together with some inspired guesswork!

CHAPTER 7 (ZIENKIEWICZ)

This is a good example of a numerical method applied to a group of practical problems, namely the stress distribution in large structures. It exploits the value of a particular tool (the electronic computer) in developing a mathematical method, the accuracy of which is dependent on how small the finite elements can be made. This is itself dependent on the size of computer available, and the numerical error so incurred is thus a measure of the coarseness of the model.

CHAPTER 8 (MURDIE)

The study of growth of population is a vast problem, for which many models may be developed. This chapter considers the specific problems relating to a single species and to two interacting species, with applications to pest control. The models highlight the problem of identifying the parameters introduced with the data from actual biological experiments.

CHAPTER 9 (DAVIES)

A differential model is developed which describes a medical condition known as *diabetes mellitus*. It is an example of a simple model which represents qualitatively broad features evident clinically whilst in no sense claiming to describe the actual biochemical mechanisms involved, which are extremely complex.

CHAPTER 10 (ASHTON)

Road traffic situations involve many features, ranging from engineering, to the social and human. This chapter concentrates on modelling the effect of fluctuations in the flow of traffic on a pedestrian wishing to cross a road, and the flow of vehicles at junctions. The variables are clearly stochastic and well-established probability distributions are tried to represent the traffic flow. Note that validation of the model with real data is considered essential.

CHAPTER 11 (LOMAS)

This provides an example of modelling in operational research, in which all significant features are present at the outset; in this case, all cash flows have to be included. Such business planning models involve decision criteria and the aim is to obtain an algorithmic process such that the insertion of specified data allows, for example, the finance manager, to arrive at an economic appraisal of a business scheme. The operation of such models often involves the use of a computer.

CHAPTER 12 (BARTHOLOMEW)

This situation concerns a policy for the control of grade structures in an organisation, by establishing rules for internal transition between grades and external movement through recruitment and wastage. It is meaningless to say whether such a model is 'true'; the point is whether it is viable. The viability may well depend on how acceptable the results are to the community on which the model is inflicted!

CHAPTER 13 (BEARD)

This is another business planning model. In insurance it is theoretically possible for a company to be ruined; the essential problem is to model the degree of risk. The variables are stochastic, but it is worth noting that successful insurance companies have found it safer to use empirical data as a basis for their probability distributions rather than using classical textbook distributions (cf. Chapter 10).

CHAPTER 14 (HILL)

This contribution concerns a practical situation in mine warfare which falls into a definite class of mathematical problems, namely the theory of games. Assumptions are made about the strategies of the contestants which essentially define the model. The recognition that the situation at hand belongs to a known class of problems for which standard theories exist is an art of modelling of great importance. It acknowledges a central *raison d'être* of applied mathematics, namely that the same mathematics may describe a wide variety of real situations which of themselves may seem entirely unrelated. (The equation for simple harmonic motion and Laplace's equation are classical examples of this). In this particular model, validation may well be difficult!

CHAPTER 15 (CARRÉ)

The interest here lies in the formal development of a network flow model in terms of an algebraic language. Problems arise in finding minimal cost flows which find immediate practical applications in the transportation of coal from mines to power stations, stock control etc.

CHAPTER 16 (ATKIN)

This is a model of an urban community in which the author claims to be able to represent all relationships within the community by means of an algebraic language. Thus the model must involve the whole situation since all relations should be included for the model to be complete. The difficulty with such models is that of assessing how far the model is representative of the situation it attempts to describe. Such assessment is likely to be very subjective, which underlines the fact that there exist models which may be judged adequate by some, but not by others!

CHAPTER 17 (CHILLINGWORTH)

This contribution is concerned with one of the more exciting recent developments in pure mathematics. The theory of catastrophe considers the conditions whereby a nominally stable state may flip to another stable state within the same overall mathematical structure by a small perturbation at points of bifurcation. Although this branch of mathematics is still in its infancy, more and more examples of real-life phenomena involving catastrophe theory are being suggested; they range over a wide variety of disciplines, from psychology to civil engineering and the social sciences.

1.4. References

ALKER, H.R. (1965). *Mathematics and Politics,* Macmillan; London
FORD, B. and HALL, G.G. (1970). 'Model Building – An Educational Philosophy for Applied Mathematics', *Int. J. Math. Educ. Sci. Technol.,* **1**, 77-83
HALL, G.G. (1963). *The Application of Mathematical Thinking,* University of Nottingham Press
HALL, G.G. (1972). 'Modelling – A Philosophy for Applied Mathematicians', *Bull. I.M.A.,* **8**, 226-228
KLAMKIN, M.S. (1970). 'On the Role of an Industrial Mathematician and its Educational Implications', Publication preprint, Ford Motor Co., USA

McLONE, R.R. (1973). *The Training of Mathematicians,* Social Science Research Council; London

MILLER, G.A. (1964). 'Mathematics and Psychology', Wiley; New York

ROSEN, R. (1967). *Optimality Principles in Biology,* Butterworths; London

SHAHANI, A.K. (1974). Private Communication

SMITH, J.M. (1968). *Mathematical Ideas in Biology,* Cambridge University Press; London

WALLIS, B.W. (1971). 'How is Mathematics Used?', *Mathematics Teaching,* No.56, 14-17

WOODS, L.C. (1969). 'What is Wrong with Applied Mathematics?', *Bull. I.M.A.,* **5,** 70

2
STEERING PROBLEMS

J.W. Craggs
Department of Mathematics, University of Southampton

[Prerequisites: none]

2.1. The problem

The problem of steering a vehicle (bicycle, car, boat) between given points is a common one, and important in quite sophisticated applications like the planning of a lunar flight or the design of a guided missile. Such steering problems belong to a much wider class of problems of control theory, in which the essential idea is to take a system from one state into another, by a route (often called a trajectory of the system) which is economically desirable and realisable in practice. An example of this wider class of problems is to plan the erection of a building at least cost.

In the simpler problems which are concerned with steering of land or sea vehicles, the car or ship may be modelled as a geometrical point, moving in two dimensions, with Cartesian coordinates (x,y) and the path or trajectory is a parametric curve given by $x = x(\lambda)$, $y = y(\lambda)$. In most cases the preferred parameter is simply elapsed time t, but in any case it should be possible to eliminate the parameter and obtain a simple equation describing the path. [In more general control problems one simply uses more coordinates $x_1(\lambda)$, $x_2(\lambda)$, $x_3(\lambda)$ ... $x_n(\lambda)$, and the motion of the system corresponds mathematically to geometric motion in a many-dimensional space.]

The solution of the steering, or control, problem consists of two steps. First, a path must be found to satisfy certain conditions as, for example:

1. That it stretch from $(x,y) = (0,0)$ to $(x,y) = (a,b)$.
2. That it lies in a given domain (a given part of the plane, as when a ship is constrained to move only on water and not on land).
3. That it minimises some economic value, which may be fuel cost, or time taken, or total cost to the operator.

The second stage of the solution is to transform it into a set of instructions to the operator, for example the path of a car is chosen by rotating the steering-wheel. If the rotation a is graduated (in degrees clockwise or anti-clockwise from the centre), the instructions may take the form 'Set $a = F(t)$, where F is tabulated against time t in the accompanying table'.

It is convenient to distinguish, from the outset, between the 'state variables', (x,y), which give the position and perhaps orientation of the system, and the 'control variables', a, which are really detailed instructions to the driver. The distinction, in the model, is that a control changes in (x,y) but (x,y) has no direct influence on a, so that the operator may change a quite arbitrarily as he chooses, within certain limits. That there are limits on a is immediately obvious in the steering-wheel problem. One cannot go on turning the wheel indefinitely in one direction!

Summing up, a mathematical model of a control problem consists of two sets of functions

$$x(t) = (x_1, x_2, ..., x_n)$$

$$\boldsymbol{a}(t) = (a_1, a_2, ..., a_m)$$

where the vectors belong to vector spaces of different dimensions, together with equations, which may be either algebraic

$$f(x, \boldsymbol{a}, t) = 0$$

or differential

$$\frac{\mathrm{d}x_i}{\mathrm{d}t} = g_i(x, \boldsymbol{a}, t) \qquad (i = 1, 2, ..., n)$$

It may seem surprising that only first-order derivatives are taken, but this is possible because one is at liberty to introduce any number of variables, x, so, for example in one dimension one may use $x_1(t)$ to denote position, $x_2(t) = \dot{x}_1(t)$ to denote speed and $x_3(t) = \dot{x}_2(t) = \ddot{x}_1(t)$ to denote acceleration.

2.2.   Restricted domains

It is convenient to illustrate the types of problem which can occur by simple examples where the basic model involves two-dimensional motion of a point. The first example shows the effect of restricting the domain in which one is allowed to steer.

EXAMPLE 1

Find the shortest path from A, $(x,y) = (-2,0)$, to B, $(x,y) = (2,0)$, in a plane, avoiding the region D: $x^2 + y^2 \leqslant 1$. (Walk past a circular lake without wetting your feet!). *See Figure 2.1.*

By definition, the shortest distance between two points in a plane is the straight line joining them. So, if P,Q are any points of the path for which the straight line PQ has no intersection with D, the path PQ will indeed be straight. Try a path ACB, where C is

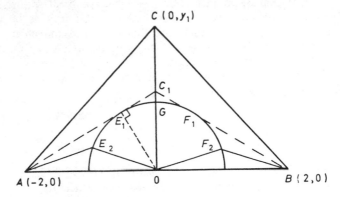

Figure 2.1   *Paths outside a circle*

$(0,y_1)$ and y_1 is large enough so that AC, CB do not intersect D. Then by Pythagoras's theorem

$$ACB = 2(4 + y_1{}^2)^{1/2}$$

and the path is shortened as y_1 is decreased. Decrease y_1 until AC touches the circle (C_1). Then $O\hat{A}C_1 = \pi/6$ rad $= 30°$. This is the best path made up of two straight lines. But now if $E_1\hat{O}C_1 = a$, $E_1C_1 + C_1F_1 = 2 \tan a$ whereas arc E_1F_1 is $2a$ and it is known that $\tan a > a$ for all a in $0 \leqslant a < \pi/2$ so the path $(AE_1;$ arc $E_1F_1; F_1B)$ is shorter than AC_1B. The stage we have now reached is that the shortest path consists of two straight lines and a circular arc. Finally try $(AE_2;$ arc $E_2F_2; F_2B)$ where $A\hat{O}E_2 = B\hat{O}F_2 = \beta$. Then the path length is $2s$ where

$$s = AE_2 + \text{arc } E_2G = \{(2 - \cos \beta)^2 + \sin^2 \beta\}^{1/2} + \pi/2 - \beta$$

Elementary calculus may be used to minimise s as β varies. The result is

$$\cos \beta = 1/2$$

$$\therefore \beta = 60°$$

The path AE_1GF_1B is therefore the shortest.

This illustrates two important theorems (*see*, for example, Craggs, 1973). A shortest path consists of arcs which are either natural shortest paths for the space, or boundary arcs of the allowed domain, and where the shortest path passes from one to the other it does so tangentially.

2.3.   Motion with a preferred direction

A second instructive problem is the well-known one of choosing the best course for a yacht in a steady wind (*see Figure 2.2*). Let the wind blow directly from the east, and consider a course at angle θ to the east. Then the speed

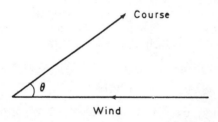

Figure 2.2   *Wind and course*

depends on θ and if θ is too near to zero, say, $-\beta < \theta < \beta$, no direct progress is possible. For $\theta > \beta$ or $\theta < -\beta$, however, the speed increases with the magnitude of θ.

Now provided that $\beta < \pi/2$, it is always possible to proceed due east, by choosing a two-leg course (tacking). *See Figure 2.3*. The problem is 'When is it better to tack than to choose a direct course?'.

EXAMPLE 2

To fix our ideas choose a model with speed v

$$v = v_0 (1 - 2 \cos \theta)$$

for $\theta > \beta = \pi/3 \text{ rad} = 60°$, where v_0 is a positive real number.

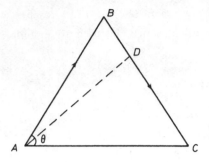

Figure 2.3 Tacking

The fastest easterly course will be given when the angle θ is chosen to maximise

$$\frac{dx}{dt} = v_0 (1 - 2 \cos \theta) \cos \theta$$

i.e. when

$$v_0 \{1/8 - 2 (\cos \theta - 1/4)^2\}$$

is maximised; i.e. when $\theta = \theta_0$, $\cos \theta_0 = 1/4$, and then $v = \frac{1}{2}v_0$.

It is certainly best to tack if the destination lies due east. But suppose the destination lies in the direction *AD*, where, say, $C\hat{A}D = \gamma > \beta$. Then the time on the tacking course is

$$\frac{AB}{\frac{1}{2}v_0} + \frac{BD}{\frac{1}{2}v_0} = \frac{AD}{\sin 2\theta_0} \left\{ \frac{\sin (\theta_0 + \gamma) + \sin (\theta_0 - \gamma)}{\frac{1}{2}v_0} \right\}$$

and the time on the direct course *AD* is $AD/v_0 (1 - 2 \cos \gamma)$. The tacking course is better whenever

$$\frac{2 \sin (\theta_0 + \gamma) + \sin (\theta_0 - \gamma)}{\sin 2\theta_0} < \frac{1}{1 - 2 \cos \gamma}$$

or equivalently, when

$$4 (1 - 2 \cos \gamma) (\sin \theta_0 \cos \gamma) < \sin 2\theta_0 \qquad (= 2 \sin \theta_0 \cos \theta_0)$$

that is, when

$$2 \cos^2 \gamma - \cos \gamma - 1/8 > 0$$

i.e.

$$\left(\cos \gamma - \frac{1 + \sqrt{2}}{4}\right) \left(\cos \gamma + \frac{\sqrt{2} - 1}{4}\right) > 0$$

But the argument only applies for $\gamma < \theta_0$, $\cos \gamma > 1/4$ so the tacking course is certainly better for

$$1/4 < \cos \gamma < (1 + \sqrt{2})/4$$

When $\gamma > \theta_0$, the course AD is shorter than $AB + BD$, and the speed is greater, so the direct course is certainly better (*see Figure 2.4*).

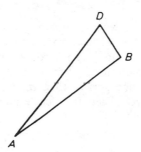

Figure 2.4   When not to tack

The final answer then, is that there is a course, $\theta = \theta_0$ which gives the best up-wind speed. For any destination up-wind of this course, the yachtsman should tack, but for any course downwind he should sail direct. This result has been proved for a special case, but applies for any reasonable relation between the speed and the course-angle.

The importance of this example is that it demonstrates that the optimum solution need not always be a smooth curve. One of the distinctions in the classical calculus of variations is the distinction between the best solution in the set of smooth curves (weak variations) and the best solution in the set of continuous but not necessarily smooth curves (strong variations). The full theory of calculus of variations enables one to deal with problems in which the speed depends on both the position and the direction of the course, as in the steering of a yacht when currents as well as wind are allowed for. However, such solutions require a rather better knowledge of calculus than is assumed in this book. Readers who are interested should read a good text on 'calculus of variations'.

2.4. Motion with restricted derivatives

The next problems in order of difficulty are those in which the rates of change of state variables are restricted in various ways. This is an important category, because in most mechanical situations one cannot control speed or position directly, but only acceleration components or possibly even higher derivatives. Problems with restricted accelerations can be very difficult, but an important sub-category consists of those problems with only a single control, in which it is the restrictions which are responsible for the existence of an optimum solution, for then the solution turns out to correspond to the boundary values of the control, and the only problem is to decide when to swing the control from one boundary to the other.

EXAMPLE 3

To push a broken-down car into a garage. Suppose one wishes to push a car in a straight line from one position to another. The simplest model assumes no resistive forces, but only a push or pull. Take the direction of motion as positive, and assume one can achieve a forward acceleration not greater than a, and a deceleration of magnitude not greater than b. Then the acceleration f is restricted by

$$-b < f < a$$

A useful device is the velocity-time diagram shown in *Figure 2.5.* The problem is to find the minimum time to move a given distance, from rest to rest.

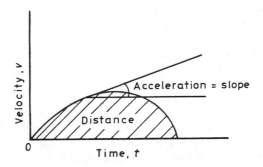

Figure 2.5   Velocity-time graph.   Shaded area = distance, slope of graph = acceleration

The natural solution is obtained by using the numerically largest available acceleration $(f = a)$ and deceleration $(f = -b)$. It is immediately clear from *Figure 2.6* that if acceleration less than a, or braking less than b, is used (dotted curve) the distance for a given time is reduced. The technique, in this solution, is to use the restrictions to calculate bounds for the domain of the variables and to look for a solution in the boundary of the domain.

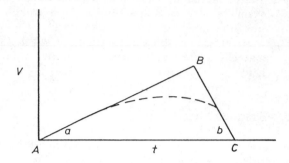

Figure 2.6   Pushing a car

EXAMPLE 4

The above model can be refined in many ways, some of which are as follows:

(1) One may allow for frictional or viscous resistance.   Then a possible equation of motion for maximum forward force is

$$m \frac{\mathrm{d}v}{\mathrm{d}t} + k_1 v + k_2 \ = \ a$$

where m is the mass, and it is the *force a* rather than the acceleration which is restricted.   The effect on the solution is simply to replace lines *AB, BC* by curves which can be calculated if one knows sufficient calculus.   The basic idea of the solution, that it corresponds to maximum acceleration, followed by maximum braking, is unchanged.

(2) In a point-to-point race, one might allow for a maximum safe speed, say $v = v_0$ as well as maximum acceleration a and braking b.   The minimum-time solution in the velocity-time diagram would then be bounded by three lines rather than two (*Figure 2.7*).

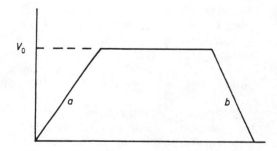

Figure 2.7   Effect of maximum speed

(3) The definition of 'optimum solution' might be changed.   For example in an economy run, a driver seeks minimum fuel consumption rather than maximum speed.   Use of brakes is then wasteful, so one prefers solutions in which the car trundles naturally to rest at the end, rather than those which require hard use of brakes.

2.5.   Manoeuvring a car

The model also applies to steering of a vehicle in two dimensions. The restriction is now on the curvature of the path, since there is a minimum diameter turning circle.   Shortest paths can be found for a point whose path has restricted curvature, as a first model for the more complicated paths of four-wheel vehicles.   Complete solutions may then require straight sections (minimal paths in the space) together with circular paths of minimum radius, but in general no other curved path should be considered.   In the following examples the radius is taken as *a*.

EXAMPLE 5

A car lies at *A*, pointing perpendicular to *AB*.   Find minimum paths, not using reverse gears, from *A* to *B*, under the assumptions stated.

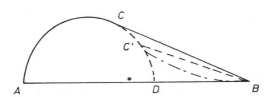

Figure 2.8   Paths for AB > 2a

(1) *AB* > 2*a* (*see Figure 2.8*)
    Clearly one begins by turning towards *B*, on the circle *ACD* of radius *a*.   Draw a tangent *CB*.   Then the required path is *ACB*.
        From Example 1, such a path as *AC'B*, even if it did not involve a corner at *C'*, would be longer and a path from *C'* to *B* incorporating a curved section to eliminate the corner (chain dotted) must be longer than *C'B*.   The result follows.
(2) *AB* < 2*a*
    If *B* lies inside *AD*, the circular path cannot go through *B*. Two possible solutions suggest themselves.
        One might drive straight *AE*, then use a circle tangent to *AE* and a tangent *FB* passing through *B* (*Figure 2.9*).

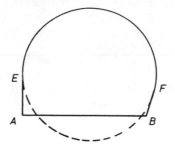

Figure 2.9   Straight and curved paths

Alternatively one might use initially a circle *AE* curving *away* from *B*, and then turn smoothly on to a circle of the opposite curvature *EB* (*Figure 2.10*).

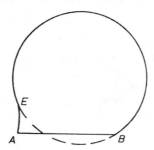

Figure 2.10   Reverse curves

It is an interesting exercise in elementary geometry to calculate the distances in the two cases.   It will be found that the second is better.

EXAMPLE 6

To drive a car from *A*, parked pointing in a direction θ_0 to *AB*, to the position *B*, pointing in a direction θ_1 to *AB* produced (car parks often ask drivers to park between given lines, *see Figure 2.11*).

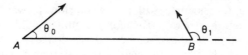

Figure 2.11   The problem

The technique is to draw both the limiting circles (radius *a*) at *A*, and both those at *B*, inserting an arrow to indicate the direction of motion (*Figure 2.12*). Now join one circle to another by a common

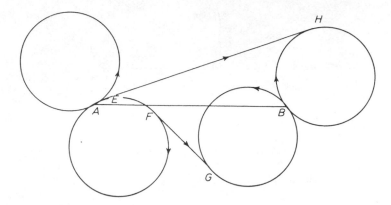

Figure 2.12 Geometrical solution

tangent, with the allowed direction at each end, e.g. *EH, FG*. Then it is generally a simple matter to work out which path gives the shorter total course.

As exercises you might like to consider (i) what happens if the car is allowed to reverse and (ii) what happens in the general case in which one or both of the circles drawn *A* intersect those drawn from *B*.

2.6. Other applications

As has been suggested at various points in the text the examples here are comparatively simple cases of the type of problem which forms the raw material of the theories of calculus of variations and of Optimal Control Theory. In pure mathematics these correspond to subsets of the advanced study of differential equations and of topological dynamics. Practical applications include the choice of minimum fuel courses for journeys to the moon or planets, minimum time courses for yachts or other sailing-ships over the oceans, and all similar transport problems. A further set of applications arises in planning the control mechanisms for automatic machines. A planing machine for example will involve a slow forward working stroke and a free return stroke. The control unit should minimise the power consumption of the machine. An automatic pottery kiln will have a carrier which moves through a heated tunnel, giving the contents specified periods at different temperatures and again the power needed (for heating) should be minimised.

2.7. Reference

CRAGGS, J.W., (1973). *Calculus of Variations,* Allen and Unwin; London

2.8. Problems for further study

1. Euclid defined a straight line as the shortest distance between two points. The logic is that you fix one end of a (fishing) line at the one point and wind it in at the other. When no more will wind in, you have the shortest distance.

  (a) If a person's waist is defined as that part of the trunk with least circumference, how would you measure it?
  (b) A 'rolling English road' has brick walls along both sides, with garden gates in them. How would you mark on the road the shortest route? What, other than a ruler or a stretched string, conveniently defines a straight line?
  (c) The edge of a pond is a trefoil, obtained by drawing three equal, symmetrically placed, circles through a given point and then digging out any part which is in any of the circles. The water area is the 'union' of the circular areas. Find the shortest closed path which encloses the pond. How would you arrange to walk along that path with your eyes open? How would you mark it with a rope, suspended between points, for a blind man?

2. List as many circumstances as you can think of in which a person with or without a vehicle will find that the speed at which he can travel depends on the direction of his motion. Example: cycling in a wind.

  (a) A skier working up-hill makes use of the fact that the friction between the snow and the ski is greater across the width of the ski than along the length. How does he get up a steep slope?
  (b) If the coefficient of friction along the ski is 0.05 and across it is 0.40, and the snow-slope is at a gradient 1 in 5, what does the shortest (distance) route up the hill look like?
  (c) If the speed at angle θ to a given line is $1 + 2\theta/\pi$ $(0 < \theta < \pi)$, what path gives the fastest progress in the direction $\theta = 0$?

3. If a vehicle can turn in any circle of radius greater than a, shortest paths will consist of circles of radius a and straight portions. If a reverse gear is not available the circular paths must be described in the 'right' direction. If a reverse gear is available a circular path may be described in either direction.

Draw three circles of equal radius *a*, with the centres *A*,*B* of two of the circles $2b = 4a \sin \theta$ apart $(0 \leqslant \theta \leqslant \pi/2)$ and draw a third circle of radius *a* touching both, as shown. Finally, draw a common tangent *CD* to the intersecting circles (on the opposite side). *See Figure 2.13.* Now work out the arc length *PQ* + *QR* + *RS* and the length *PC* + *CD* + *DS*, where *P*,*S* are on the common diameter *AB*. Which is greater? Does it depend on θ? Can they ever be equal? Next, compare the arc length *LR* + *RQ* + *QM* with the length *CD* + *DC* + *CM*. Now answer the following questions.

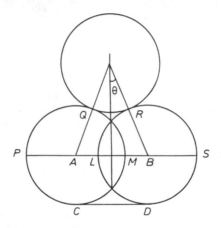

Figure 2.13   Problem (3)

(a) A car parked at *P* facing up the page is to be left at *S* facing down the page. What is the shortest path?

(b) A car parked at *L* facing up the page is to be parked at *M* facing down the page. What is the shortest path (i) if reverse gear is available, (ii) if reverse gear is not available?

(c) A car parked at *C* facing up the page is to be left at *P* facing up the page. What is the shortest path (i) if reverse gear is available, (ii) if reverse gear is not available?

4. It is required to provide transport on a given planet. The surface is so irregular that wheeled or tracked vehicles are impracticable. There is no atmosphere. It is therefore proposed that the vehicle be provided with steerable rocket motors which may be directed at any angle to the vertical to provide lift and horizontal acceleration. The fuel consumption of a rocket (per unit time) is approximately proportional to the square of the thrust (why?). Discuss the problem of finding a minimum fuel path between two points on the surface of the planet.

5. The power output of an electric motor is proportional to the square of the current consumed but the current can be reduced (on a constant voltage supply) only by using a resistance which itself consumes power proportional to the difference between the maximum permissible current and the current used.   The motor is used to drive a vehicle at an average speed V against a resistance F.   Discuss the power consumed:

(a) If a resistance is used to control the current so as to achieve a constant speed.
(b) If the current is rapidly switched on and off so that it is on for a given fraction of each second.

Which system gives a more economical control of the vehicle?

3
WHY BUILD THREE-STAGE ROCKETS?

B. Noble
University of Wisconsin and University of Oxford

[Prerequisites: elementary mechanics and ordinary differential equations]

3.1. Identification of the problem

A *model* of a rocket is a simplified version of the real thing.   The
essence of modelling is that the model must resemble the original in
one or more essential characteristics.   For instance, when modelling a
rocket, we might be interested in making the model look like a real
rocket (but then it need not have an engine that works), or we might
concentrate on making a working model rocket engine (in which case
the appearance of the rocket is of less importance), and so on.   We
shall be concerned with a conceptual model of a rocket.   The object
of the model is to enable us to understand why real rockets are constructed in a certain way.   It is well known that to place a satellite
in orbit, we use a rocket engine that is built in three stages.   The
first stage is fired; when the fuel in this stage is exhausted, the fuel
container is jettisoned.   Similarly with the second stage; and finally
the third stage.   We wish to understand why we make three-stage
rockets and not one- two- or four-stage rockets, say.

The most difficult step in mathematical modelling is the first one.
We have to decide which factors are important and which are not.
A rocket is a very complicated system.   It must have a powerful
engine, a strong construction, low air-resistance, and so on.   The key
requirement is clearly that the engine must be powerful enough to
accelerate the rocket to a sufficiently high speed.   If the final speed
is too low, the rocket will simply fall back to earth.

A red herring here, from our point of view, would be to consider
at this point how a satellite is placed in orbit — the rocket carrying

the satellite is fired vertically from rest on the surface of the earth, and turned over in flight into an orbit moving round the earth. However, the main thing that we are interested in is that the final velocity of the rocket is large enough. This leads us to our first question:

    1. How fast does a satellite move in orbit?

We must next consider how we can achieve this velocity. In any dynamical problem, our first thought is to try to apply Newton's laws of motion. Newton's second law for a fixed mass says that

$$\text{force} \;=\; \text{mass} \times \text{acceleration}$$

To reach a large final velocity in a reasonable time we need a large acceleration, i.e. a large force and a small mass. This leads us to two more questions:

    2. What determines the force exerted by a rocket engine?
    3. What determines the mass of a rocket-satellite system?

 · When these questions are answered quantitatively we will find that it is impractical to build a one-stage rocket to orbit a satellite.

3.2. The motion of the satellite

Let us consider Question (1). It turns out that the key factor is the force of gravity. Qualitatively we visualise the situation shown in *Figure 3.1*, where the full line *C* denotes the surface of the earth, and the dotted line represents the path of the satellite. If no forces were acting on the satellite at the point *S*, it would move along the straight line *ST* tangent to its path (Newton's first law of motion). In fact, the pull of gravity, acting along the line *SO* towards the centre of the earth, will make the satellite move in an approximately circular path.

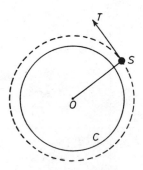

Figure 3.1

We can simplify the situation by assuming that the earth is fixed in space, spherical, and has a spherically-symmetric density distribution. Newton's law of gravitation says that every particle of matter in the universe attracts every other particle according to the inverse of the distance squared. It is a well-known consequence of this law that the gravitational force exerted by a sphere of uniform density on a mass outside the sphere is the same as if the mass of the sphere is concentrated at its centre (*see* Problem 1). In the simple model of the earth's gravitational field, the force exerted by gravity on a mass m at a distance r from the centre of the earth (where r is greater than the radius of the earth) is given by

$$\frac{km}{r^2} \qquad (k = \text{constant})$$

and this force is directed towards the centre of the earth. To determine the value of k we note that, if g denotes the acceleration due to gravity at the surface of the earth, Newton's second law of motion for a fixed mass gives

$$mg \;=\; \text{force due to gravity} \;=\; \frac{km}{R^2}$$

where R is the radius of the earth. Hence $k = gR^2$, and the gravitational force on a satellite of mass m at a distance r from the centre of the earth is given by

$$G \;=\; mg\left(\frac{R}{r}\right)^2$$

The next step is to write down the equations of motion of a satellite moving in a plane through the centre of our idealised spherical earth. If we choose fixed x- and y-axes in this plane, as shown in *Figure 3.2*, we have

$$m\ddot{x} \;=\; -G \cos \theta$$
$$m\ddot{y} \;=\; -G \sin \theta$$

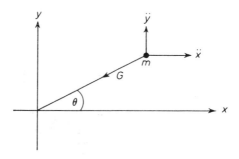

Figure 3.2

Putting $\cos \theta = x/r$, $\sin \theta = y/r$, then for a *circular* orbit ($r =$ constant) we can integrate directly to obtain

$$v^2 \;=\; g\,\frac{R^2}{r}$$

where
$\quad v^2 \;=\; \dot{x}^2 + \dot{y}^2$
$\qquad g \;=\;$ acceleration due to gravity
$\qquad R \;=\;$ radius of the earth
$\qquad r \;=\;$ radius of the satellite's orbit

We note that $g = 9.81$ m/s^2 and $R = 6400$ km can be found from measurements made on the earth's surface. Also v is independent of the mass of the satellite, and it will be roughly independent of r for orbits less than a few hundred miles above the surface of the earth. As a practical answer, suppose that the satellite is moving in an orbit 600 km above the surface of the earth. Then

$$v \;=\; R\left(\frac{g}{r}\right)^{1/2}$$

$$\approx 7.6 \text{ km/s}$$

This answers our first question.

3.3. The force exerted by a rocket engine

We next consider the second question: What can we say about the force exerted by a rocket engine? A simple model of a rocket consists of a rocket engine and a container for the fuel. The fuel is burned in the engine. The resulting gases are expelled from the end of the rocket. The backward streaming gases exert a forward thrust on the rocket.

Insight into the action of a rocket engine can be obtained from the principle of conservation of momentum. The momentum of a single particle is defined as its mass multiplied by its velocity. To find the total momentum of a system we add together the momenta of the separate parts of the system. The law of conservation of momentum states that in a closed system on which no external forces are acting, the total momentum of the system is constant independent of time.

We can use this principle to obtain some understanding of the action of a rocket engine by considering the simplified model illustrated diagrammatically in *Figure 3.3*. We visualise the rocket moving in a straight line, propelled by gases being expelled behind it. (We neglect other forces such as gravity, air resistance and so on.) Suppose that at time t the rocket has a mass $m(t)$ and velocity $v(t)$. At time $t + \Delta t$ the rocket has a mass $m(t + \Delta t)$. The loss of mass is given by

$$-\{m(t + \Delta t) - m(t)\} = -\frac{dm}{dt}\Delta t + 0(\Delta t^2)$$

using Taylor's theorem.   In our simplified model we assume that this

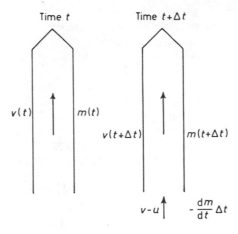

Time t Time $t+\Delta t$

$v(t)$ $m(t)$

$v(t+\Delta t)$ $m(t+\Delta t)$

$v-u$ $-\frac{dm}{dt}\Delta t$

Figure 3.3

loss of mass is that of gas ejected at a velocity relative to the rocket, u, which is constant.   This means that the velocity of the gas relative to the earth is given by $v(t) - u$.

We can apply the principle of the conservation of momentum to the system shown in *Figure 3.3*.   This gives

$$m(t)v(t) = m(t + \Delta t)v(t + \Delta t) - \left(\frac{dm}{dt}\Delta t\right)(v(t) - u) + 0(\Delta t^2)$$

i.e.
| momentum of rocket at time t | = | momentum of rocket at time $t + \Delta t$ | + | momentum imparted to gas at time $t + \Delta t$ |

(The gases emitted before time t would simply add the same terms to both sides of this equation and can therefore be ignored.)

Using Taylor's theorem to expand $m(t + \Delta t)$, and taking the limit as $\Delta t \to 0$, we obtain

$$m\frac{dv}{dt} = -\frac{dm}{dt}u \qquad (1)$$

This is an instructive formula.   The left-hand side represents the inertial force on the rocket.   Hence the thrust of the rocket due to expelled gas, which we denote by T, is given by

$$T = -\frac{dm}{dt}u$$

In words, the thrust is equal to the product of the rate at which the fuel is being used up and the velocity of the ejected gases relative to the rocket. This has answered our second question.

The differential Equation (1) can be written:

$$\frac{dv}{dt} = - u \frac{d (\ln m)}{dt}$$

Integrating for u = constant, gives

$$v(t) = v_0 + u \ln \left(\frac{m_0}{m(t)} \right) \tag{2}$$

where m_0 is the initial mass of the rocket and v_0 is its velocity at time $t = 0$. This simple result says that the change in velocity of the rocket depends on only two things:

(a) The (assumed constant) velocity u of the expelled gases relative to the rocket.

(b) The ratio of the mass of the rocket at time $t = 0$ to the mass at time t.

3.4. The mass of a rocket-satellite system

We now come to the third question: What determines the mass of a rocket-satellite system?

Consider a rocket of mass m_0 consisting of

(a) The 'payload', of mass m_P

(b) The fuel, of mass m_F

(c) The structural weight of fuel container and engines, m_S

First consider a simple model in which all the fuel is expelled, leaving a final mass $m_P + m_S$. The fuel container and engines are then jettisoned, leaving the payload moving at a velocity given by Equation (2):

$$v = u \ln \left(\frac{m_0}{m_P + m_S} \right) \tag{3}$$

In practice it is difficult to build engines and fuel containers with a total mass less than about an eighth or a tenth of the weight of the fuel. We introduce a constant λ defined by

$$m_S = \lambda (m_F + m_S) = \lambda (m_0 - m_P)$$

i.e. the structure is a fraction λ of the combined mass of the fuel plus structure. Then Equation (3) gives:

$$v = u \ln \left(\frac{m_0}{\lambda m_0 + (1 - \lambda) \, m_\mathrm{P}} \right)$$

This leads immediately to an important result. For a given u, the maximum velocity that can be reached by the rocket will be achieved when the payload is zero (as we should expect) and this maximum velocity is given by

$$v = u \ln \left(\frac{1}{\lambda} \right)$$

For modern fuels, a representative value of u is about 3 km/s. If $\lambda = 0.1$, then

$$v = 7 \text{ km/s}$$

Since a satellite in orbit moves at about 7.6 km/s, and our maximum velocity has been obtained by ignoring air-resistance, gravity, and by carrying no payload at all, it is clear that a rocket of this type cannot be used to orbit a satellite.

One reason for this poor performance is the fact that the engine has to accelerate the whole dead weight of the structure up to the final velocity. When the fuel is almost exhausted, the engine is expending most of its effort on accelerating an almost empty fuel tank. It would be more efficient if we jettison useless weight as the burning proceeds. Consider the idealised situation illustrated diagrammatically in *Figure 3.4*. Between times t and $t + \Delta t$ we assumed that a fraction λ of the change in mass is structural mass which is simply thrown away, and a fraction $(1 - \lambda)$ is burned and expelled as gas at velocity u. (It would of course be impossible to build a

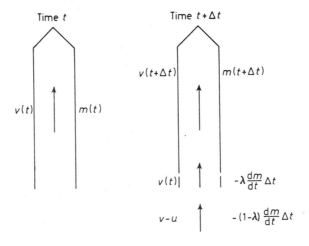

Figure 3.4

rocket that worked in this way, but such an objection is irrelevant since we are predicting ideal performance. If the ideal gives us what we want we can then try to approximate the ideal in practice.)

Conservation of momentum now gives

$$m(t)v(t) = m(t + \Delta t)v(t + \Delta t) - \lambda \frac{dm}{dt} \Delta t\, v(t) - (1 - \lambda)\frac{dm}{dt}\Delta t(v - u)$$

Rearranging and letting Δt tend to zero, we find

$$m \frac{dv}{dt} = (1 - \lambda)\, u\, \frac{dm}{dt}$$

Integrating

$$v(t) = (1 - \lambda)\, u\, \ln\left(\frac{m_0}{m(t)}\right)$$

This is similar in form to our previous formula, but there is one significant difference, namely that the final mass, when the fuel is used up, is simply the payload, since the structural weight has been jettisoned as the fuel was burned. Hence the final velocity is

$$v = (1 - \lambda)\, u\, \ln\left(\frac{m_0}{m_P}\right)$$

If we start with a given u, λ and initial mass m_0, a payload can be accelerated up to any velocity we like, although the greater the required speed, the smaller the payload that can be accelerated up to this speed.

Suppose that to take account of factors like air resistance, gravity, and so on, our idealised rocket must be designed to reach a final speed of 10.5 km/s (instead of 7.6 km/s, as before) and we take $\lambda = 0.1$, $u = 3$ km/s, as before. Then

$$\frac{m_0}{m_P} \approx 50$$

i.e. the payload is one-fiftieth of the initial mass of the whole system. For a one tonne payload we need a fifty tonne rocket.

3.5. Practical realisation of the ideal performance

The next question is: How can we approximate this ideal performance in practice? The answer is to build the rocket in stages and shed each stage when the fuel in it is used up. Let

$$m_i = \text{weight of (fuel + structure) in the } i\text{th stage,}$$

and suppose that λm_i of this is structure and $(1 - \lambda)m_i$ is fuel, where for simplicity we assume that λ is the same for all stages.

Suppose also for simplicity that the velocity u of the exhaust gases is the same for all the stages.

We analyse the performance of a three-stage rocket. The initial mass of the rocket is given by

$$m_0 = m_P + m_1 + m_2 + m_3$$

When the first stage fuel is exhausted, the mass left is

$$m_P + \lambda m_1 + m_2 + m_3$$

and the velocity is (from our first model)

$$v_1 = u \ln \left(\frac{m_0}{m_P + \lambda m_1 + m_2 + m_3} \right)$$

The structural mass m_1 is then dropped, and the second stage is fired. The velocity when the fuel in the second stage fuel is exhausted is

$$v_2 = v_1 + u \ln \left(\frac{m_P + m_2 + m_3}{m_P + \lambda m_2 + m_3} \right)$$

Similarly the velocity when the third stage fuel is exhausted is

$$v = v_2 + u \ln \left(\frac{m_P + m_3}{m_P + \lambda m_3} \right)$$

We are now in a position to determine the sizes of the three stages so as to give the maximum payload. More precisely, suppose that we know the final velocity v that we wish to achieve, the velocity u of the exhaust gases relative to the rocket, and the structural factor λ. For a given total initial mass m_0, how do we choose m_1, m_2, and m_3, so as to maximise the payload m_P?

The problem is: For fixed m_0, v, u, λ, maximise m_P subject to

$$m_P + m_1 + m_2 + m_3 = m_0$$

and

$$\frac{v}{u} = \ln \left(\frac{m_0}{m_P + \lambda m_1 + m_2 + m_3} \right) \left(\frac{m_P + m_2 + m_3}{m_P + \lambda m_2 + m_3} \right) \left(\frac{m_P + m_3}{m_P + \lambda m_3} \right)$$

This problem is straightforward in principle but can become messy in detail unless we simplify the algebra. An easy way to solve the problem is to introduce new variables

$$a_1 = \frac{m_0}{m_P + m_2 + m_3} , \quad a_2 = \frac{m_P + m_2 + m_3}{m_P + m_3}, \quad a_3 = \frac{m_P + m_3}{m_P}$$

The problem can then be reduced to the following: Minimise $a_1 a_2 a_3$ subject to

$$\frac{v}{u} = \ln\left\{\left(\frac{a_1}{1 + \lambda\,(a_1 - 1)}\right)\left(\frac{a_2}{1 + \lambda\,(a_2 - 1)}\right)\left(\frac{a_3}{1 + \lambda\,(a_3 - 1)}\right)\right\}$$

(*See* Problem 3.)

The nice thing about this formulation is that it is symmetric in a_1, a_2, a_3, which tells us immediately that the optimum must be attained when $a_1 = a_2 = a_3$. If we call this common value a, we easily see that

$$\frac{a}{1 + \lambda\,(a - 1)} = \exp\left(\frac{v}{3u}\right)$$

or

$$a = \frac{1 - \lambda}{P - \lambda} \qquad \text{where } P = \exp\left(-\frac{v}{3u}\right)$$

We recall that the quantity we are optimising is $a_1 a_2 a_3$, which is the ratio m_0/m_P, so that the maximum payload is given by the formula

$$\frac{m_0}{m_P} = \left(\frac{1 - \lambda}{P - \lambda}\right)^3 \qquad \text{where } P = \exp\left(-\frac{v}{3u}\right)$$

For the case considered previously at the end of Section 3.4, $v/u = 3.5$, $\lambda = 0.9$, and we find

$$\frac{m_0}{m_P} = 77$$

This means that for a one tonne payload we need a 77 tonne three-stage rocket. This can be contrasted with the 50 tonnes that we found previously for the ideal rocket that sheds its structure continuously.

We wish to investigate next whether it would pay to build a two- or four-stage rocket instead of a three-stage rocket. From the analysis of the three-stage rocket, it will be clear that the optimum ratio of m_0/m_P for an n-stage rocket, assuming that u and λ are the same for all stages, is given by

$$\frac{m_0}{m_p} = \left(\frac{1 - \lambda}{P - \lambda}\right)^n \qquad \text{where } P = \exp\left(-\frac{v}{nu}\right)$$

As a check, if we let n tend to infinity, this tends to the result found previously for the ideal rocket in which the structure was shed continuously:

$$\frac{m_0}{m_p} = \exp\left(\frac{v}{(1 - \lambda)\, u}\right)$$

We can now make a table of the mass of an n-stage rocket to orbit a one tonne satellite, for the v/u and λ assumed previously:

n = no. of stages	1	2	3	4	5	∞
mass (tonnes)	-	149	77	65	60	50

It is clearly worthwhile to use three stages instead of two. However, the weight reduction obtained by using four instead of three is counterbalanced by the complexity and expense of making an extra engine and fuel tank. In practice three stages give the optimum design.

Our analysis so far does not seem to depend on the size of the rocket engine, yet it is clear that we must have a large engine to lift a 77 tonne rocket off the ground. One of the ingenious things about our analysis is that we have been able to ignore gravity when considering the rocket engine that we need. However, at this point we must consider the process by which the satellite is placed in orbit. The rocket is initially at rest on the surface of the earth, and it is fired vertically upwards. This means that the engine must be powerful enough to overcome the gravitational force on the rocket. We have seen earlier that the thrust of the rocket is given by

$$T = - \frac{dm}{dt}\, u$$

Current practice is to make the thrust T at lift-off about 1.25 times the weight of the rocket. Also a typical value for the velocity u of the gases from the engine is 3 km/s. If m is in tonnes the rate at which the rocket engine must burn is given by

$$- \frac{dm}{dt} = \frac{1.25\, mg}{u} = \frac{(1.25)(9.81)(60)}{3000}\, m = 0.24\, m \text{ tonnes/min}$$

The rocket engine must burn up about one-quarter of the total mass of the rocket per minute. We can see why rocket engines fire for only a few minutes. The size of the engine required to lift our 77 tonne rocket off the ground can be appreciated from the fact that the above formula says that it must burn fuel at the rate of about 20 tonnes/min.

Of course the discussion given here is only the start of the story. Some of the complexities involved in the mathematical modelling of real rockets are discussed in Ball and Osborne (1967). This is a delightful book, which illustrates beautifully the development of more complicated models to explain various aspects of space vehicle performance. It is recommended for the serious student rather than the casual reader.

Reference

BALL, K.J. and OSBORNE, G.F. (1967). *Space Vehicle Dynamics,* Oxford University Press; London

3.7. Problems for further study

1. Deduce the velocity of a satellite in orbit from the following model. Assume that the earth is fixed in space, spherical, and of spherically symmetric density. Newtons' law of gravitation says that every particle of matter attracts every other particle according to the inverse of the distance between the particles squared, and the force of attraction is proportional to the product of the masses. Deduce that the gravitational force on the satellite is the same as if the mass of the earth were concentrated at its centre. Write down the equations of motion of the satellite moving in a plane through the centre of the earth, in a circular orbit whose centre is the centre of the earth. Deduce that

$$v^2 = \frac{gR^2}{r}$$

where v = velocity of satellite, g = acceleration due to gravity, R = radius of the earth, and r = radius of the satellite's orbit. Deduce that the velocity of a satellite in an orbit of a few hundred kilometres above the earth is approximately 7.6 km/s or 27,000 km/h.

2. Derive the principle of conservation of momentum by the following argument. Suppose that we have a system of n particles, the mass of the ith particle being m_i and its position at time t being denoted by (x_i, y_i, z_i). Assume that the system is isolated, i.e. there are no external forces. The only forces acting are the action of the jth particles on the ith, and this we denote by $(F_{xij}, F_{yij}, F_{zij})$. (Note that, whatever the interactions between the particles, the force exerted by the jth particle on the ith is exactly equal to the force exerted by the ith on the jth. This is Newton's third law — action and reaction are equal and opposite.)

3. Work out the formula for the payload that can be put into orbit by an n-stage satellite. Verify the limiting form as $n \to \infty$. Derive this directly from a model in which the casing weight is shed continuously. Verify the weights of rocket required to orbit a one tonne satellite as given in the table.

4. Prove the key result required to derive the formula for the payload that can be put into orbit by an n-stage satellite, namely that if w_r = total weight of the rocket when the engines of the rth stage are fired, then, in the optimum design that will place maximum payload in

orbit, the ratio w_{r-1}/w_r is a constant independent of r for
$r = 1, 2, ..., n$ (w_n is equal to the weight of the payload).

5.   Since our formulae depend only on the ratio m_0/m_P, it would
seem that a 77 kg rocket could put a 1 kg satellite in orbit round
the earth.   This is clearly not correct.   Give at least two reasons why
the model predicts this incorrect result.   List any other weak points
about our rocket-satellite model.   Which of them are most important?
How could our model be improved to deal with some of the factors
mentioned.

6.   How would you find the masses of each of the stages of a three-
stage rocket to orbit maximum payload if u and λ are constants, but
different for each of the three stages?

4

LIQUID FLOWING FROM A CONTAINER

N. Curle
Department of Applied Mathematics, University of St. Andrews

[Prerequisites: elementary fluid mechanics and elementary ordinary differential equations]

4.1.   Introduction

A readily visualised everyday problem arises from the practice of supplying milk in polythene sachets.   To transfer the milk into a jug the sachet is first placed in a 'dispenser', thus ensuring that the sachet retains an essentially fixed shape.   A small hole is cut at each of the two free corners of the sachet, one for pouring out the milk, the other to allow air to enter the sachet as the milk leaves.   The purpose of the exercise is to calculate how long it takes for the milk to be poured out from the sachet.

4.2.   The Helmholtz-Kirchoff free streamline theory

The model to be used is based upon the ideas and results of the Helmholtz-Kirchoff free streamline theory (Milne-Thomson, 1968; Chapters 11 and 12).   Consider a large container which encloses a fluid of density ρ at a pressure p_1.   Outside the container is a much lighter fluid, at an ambient pressure p_0, where $p_0 < p_1$.   If a small slit is opened in the wall of the container, fluid flows into the region of lower pressure, and a jet forms, having a free boundary between itself and the surrounding (lighter) fluid.   When a steady flow has been set up, the speed u of the jet may be derived as follows.

Consider a streamline such as AB in *Figure 4.1*.   In steady flow, and when gravity is ignored, Bernoulli's equation states that $p + \frac{1}{2}\rho u^2$ is constant along a streamline, so

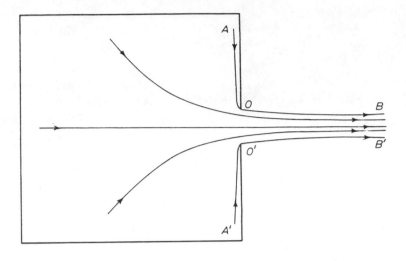

Figure 4.1   Flow from a closed container

$$p_A + \tfrac{1}{2}\rho u_A{}^2 \;=\; p_B + \tfrac{1}{2}\rho u_B{}^2 \qquad\qquad (1)$$

Now mass is conserved in any given stream tube, and since the streamlines in the vicinity of B are very much closer together than in the vicinity of A, it follows that

$$u_B \gg u_A$$

Accordingly Equation (1) may be written as

$$p_A \simeq p_B + \tfrac{1}{2}\rho u_B{}^2$$

Now the pressure must be continuous across the free surface of the jet, so

$$p_B = p_0$$

Likewise, since fluid elements near to A are virtually at rest,

$$p_A = p_1$$

Thus

$$u_B{}^2 = \frac{2\,(p_1 - p_0)}{\rho}$$

The volume of fluid flowing out of the container in unit time is

$$V = u_B A_B$$

where A_B is the cross-sectional area of the fully developed jet, in the vicinity of B. The area A_B is, in general, not equal to the area A_0 of the hole, since the jet tends to contract. The reason for this contraction is the inability of the fluid to change direction instantaneously near O and O', and the extent of the contraction effect increases with the angle through which the streamlines ultimately have to turn. The contraction ratio

$$C = A_B/A_0$$

accordingly depends on the angle a which either side of the container makes with the axis of symmetry.

When the flow is two dimensional, the problem may be solved exactly using complex variable theory, and calculated contraction ratios for various angles a are shown in *Table 4.1*, together with some experimentally observed values. It will be noted that the experimental values are slightly higher than the theory predicts, the discrepancy being between 1% and 8%.

Table 4.1   Contraction ratios

a	0	$22.5°$	$45°$	$67.5°$	$90°$	$180°$
C (theory)	1	0.855	0.745	0.666	0.611	0.500
C (expt)	-	0.882	0.753	0.684	0.632	0.541

The situation in three-dimensional axi-symmetric flow, through a circular hole, is just a little different. It may be shown (*see* Milne-Thomson, 1968; Birkhoff and Zarantonello, 1957) that $C = \frac{1}{2}$ when $a = 180°$, and clearly $C = 1$ when $a = 0$ (since the streamlines are all straight lines in this case). When $a = 90°$, the case shown in *Figure 4.1*, an extremely elegant analysis by Garabedian (1956) has shown that the contraction ratio is almost certainly 0.58 to two significant figures. Given that the experimental values in *Table 4.1* for two-dimensional flow are all slightly higher than the theoretical values, and that the theoretical values for axi-symmetric flow are slightly less than those for two-dimensional flow, we will probably not go far wrong if we use the *Table 4.1* theory values hereafter.

Consider now the situation of a container which is open at both 'ends', with gravity taken into account. For example, we take a funnel of semi-angle a, as shown in *Figure 4.2*. Now Bernoulli's equation states that $p + \frac{1}{2}\rho u^2 + \rho g y$ is constant along a streamline in steady flow. Hence, for the streamline AB, we have

$$p_B + \tfrac{1}{2}\rho u_B{}^2 = p_A + \tfrac{1}{2}\rho u_A{}^2 + \rho g h \qquad (2)$$

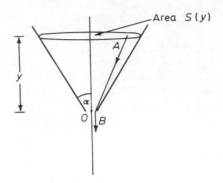

Area $S(y)$

Figure 4.2 Flow from an open container

where h is the height of the surface of the liquid above OB. (We note that h will decrease as fluid flows from the funnel, but it will decrease very slowly if the exit area is small enough, in which case the flow will be approximately steady.)

The pressures at A and B must both be equal to the external pressure p_0. Again, as before, u_A^2 is negligible by comparison with u_B^2. Thus Equation (2) becomes

$$\tfrac{1}{2}\rho u_B^2 \;=\; \rho g h$$

or

$$u_B^2 \;=\; 2gh$$

4.3. Emptying time for arbitrary fixed container

Consider a container which initially contains fluid to a height h above a small hole, through which the fluid begins to escape at time $t = 0$. For simplicity we take the container to be fixed in position throughout.

Suppose the height of the surface is y at time t, and that the cross-sectional area of the surface is $S(y)$. In time Δt, the level falls by an amount $-\Delta y$. Then the volume of fluid leaving the container is

$$-S(y)\ \Delta y$$

But the speed at which the fluid flows from the container is $(2gy)^{1/2}$, so the volume of fluid leaving in time Δt is

$$A(2gy)^{1/2}\ \Delta t$$

where A is the cross-sectional area of the exit and C is the contraction coefficient. Equating these two expressions, we have

$$-S(y) \frac{dy}{dt} = AC(2gy)^{1/2} \tag{3}$$

This is a simple separable first-order, first-degree equation, from which y may be deduced as a function of t. In fact

$$AC(2g)^{1/2}t = -\int_h^y \frac{S(y)}{y^{1/2}} \, dy \tag{4}$$

after using the initial condition that $y = h$ when $t = 0$.

As a specific example, consider a conical container having a semi-angle a. Then the cross-sectional radius at a height y is $y \tan a$, and the cross-sectional area is

$$S(y) = \pi (y \tan a)^2$$

Equation (4) thus becomes

$$AC(2g)^{1/2}t = -\pi \tan^2 a \int_h^y y^{3/2} \, dy$$

$$= \frac{2}{5} \pi \tan^2 a(h^{5/2} - y^{5/2})$$

The container is empty when $y = 0$, and the time taken to achieve this position is

$$t = \frac{\frac{2}{5} \pi \tan^2 a \, h^{5/2}}{AC(2g)^{1/2}} \tag{5}$$

By suitably re-writing this expression, (*see* Problem (1)), it may be shown that the mean speed of efflux is $5/6 \, (2gh)^{1/2}$. This is much closer to the maximum efflux speed, $(2gh)^{1/2}$, than to the minimum value, zero, since most of the fluid flows out whilst y is close to h, and $(2gy)^{1/2}$ is close to $(2gh)^{1/2}$. The precise value of the constant, 5/6 in this case, will depend upon the shape of the container, and should be greater than 1/2 if the container is fatter at the top than at the bottom; this can readily be verified by detailed calculations. Likewise, for a cylindrical container, where the cross-sectional area does not vary with height y, we may anticipate that the constant will be precisely 1/2.

The result of Problem (2), which deals with the family of containers whose shapes are defined by the cross-sectional areas

$$S(y) = S_0 \, (y_0/h)^{2n}$$

is that such a container will be empty after a time

$$t = \frac{V_0}{AC\bar{v}}$$

where V_0 is the initial volume of fluid in the container, and the mean velocity of efflux is

$$\bar{v} = \frac{2n + \frac{1}{2}}{2n + 1}(2gh)^{1/2}$$

It may be noted that $\bar{v} = \frac{1}{2}(2gh)^{1/2}$, as anticipated, for a cylindrical container $(n = 0)$, and that $\bar{v} = \frac{5}{6}(2gh)^{1/2}$ for a conical container $(n = 1)$.

4.4.   Application to milk sachet

Milk is delivered to the author's family in sachets, shaped roughly as shown in *Figure 4.3*.   The milk pours through a hole at B, and air enters through a hole at A.   Because of this latter hole there is a maximum angle through which the sachet can be tilted.   This maximum value of θ is initially approximately $10°$, although the angle can be increased as the level of the surface falls.

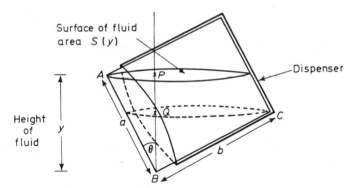

Figure 4.3   Polythene milk sachet: initial stages of pouring

As a first attempt at a model, let us take θ to be the constant value a (= $10°$).   The volume initially occupied by the milk consists of two portions, the lower portion taking the form of a sheared cone, apex downwards, of height $BQ = h_1 = b \sin a$, and a roughly elliptical base of area S_0 say.   The upper portion is a sheared cylinder of height $PQ = h - h_1$ with the same roughly elliptical cross-section of area S_0.   Thus

$$S(y) = S_0 (y/h_1)^2, \qquad 0 \leqslant y \leqslant h_1$$
$$= S_0, \qquad\qquad h_1 \leqslant y \leqslant h \tag{6}$$

The time taken to empty the milk is then determined from Equation (4) as

$$AC (2g)^{1/2} t = \int_0^h \frac{S(y)}{y^{1/2}} \, dy$$

$$= \int_0^{h_1} S_0 \left(\frac{y}{h_1}\right)^2 \frac{dy}{y^{1/2}} + \int_{h_1}^h S_0 \frac{dy}{y^{1/2}}$$

$$= 2S_0 (h^{1/2} - \frac{4}{5} h_1^{1/2}) \tag{7}$$

It may readily be verified that this result agrees with those previously obtained for a cylindrical container · $(h_1 \rightarrow 0)$ and for a conical container $(h_1 \rightarrow h)$.

Equation (7) may be re-written as

$$t = \frac{2S_0 h \{1 - \frac{4}{5} (h_1/h)^{1/2}\}}{AC(2gh)^{1/2}} \tag{8}$$

Now the cross-sectional area S_0 is related to h_1 and h by the condition that the initial volume of milk, V_0, is precisely one pint. Thus

$$V_0 = \frac{1}{3} S_0 h_1 + S_0 (h - h_1)$$

$$= S_0 (h - \frac{2}{3} h_1)$$

and so Equation (8) becomes

$$t = \frac{2V_0}{AC(2gh)^{1/2}} \frac{1 - \frac{4}{5} \left(\frac{h_1}{h}\right)^{1/2}}{1 - \frac{2}{3} \frac{h_1}{h}} \tag{9}$$

In an experiment carried out by the author, the sachet had dimensions $a = 12.7$ cm, $b = 15.2$ cm, and the maximum initial angle of tilt was $a = 10°$. Thus

$$\frac{h_1}{h} = \frac{b}{a} \tan a = 0.212$$

$$\left(\frac{h_1}{h}\right)^{1/2} = 0.460$$

$$h = (12.7 \cos 10°) \text{ cm} = 12.51 \text{ cm}$$

and hence

$$(2gh)^{1/2} = 157 \text{ cm/s}$$

Upon substituting these values into Equation (9) it follows that

$$t = \frac{0.00938 \ V_0}{AC} \tag{10}$$

Upon further substituting the value

$$V_0 = 1 \text{ pint} = 0.568 \text{ litre}$$

and the estimated area A through which the milk poured as

$$A = 0.258 \text{ cm}^2$$

Equation (10) finally yields

$$t = 27.7 \text{ s}$$

using the value $C = 0.745$ as the contraction ratio for a (maximum) semi-angle of 45°.

As far as could be ascertained without a stop watch, the milk emptied from the sachet in 25 s, so the agreement between theory and experiment is reasonable. However, it is good to reme.nber that even a poor theory can agree well with experiment if there is a for-tuitous cancellation of effects which the theory neglects. We shall accordingly consider possible improvements to this theory.

4.5. The effects of the starting period

The theory presented above is a steady-state theory, which assumes that the efflux velocity is precisely $(2gy)^{1/2}$ when the head of fluid is of height y. Clearly the efflux velocity must be zero initially, so the conditions of the theory will only be reached after a sufficiently large time. The theory will accordingly underestimate the time taken to empty the container.

Relatively little is known about the starting-up of free-boundary flows, but fortunately the relevant information for our needs is avail-able. For the case of a jet issuing from a container whose sides con-tain an angle $2\alpha = \pi$, it has been shown (Curle, 1956) that there are solutions for which the unsteadiness decays with time like exp $(-\lambda t)$, where the smallest value of λ is

$$\lambda = \frac{(\pi + 2)\ U}{2d}\ (2 - \sqrt{2}) = \frac{1.506\ U}{d}$$

In this equation U is the speed of the jet in the steady state, and $2d$ is the width of the (two-dimensional) slit. For axi-symmetric flow, with a circular hole of radius a, we may accordingly assume that the appropriate value of λ is of order

$$\lambda = \frac{1.506\ U}{a} \simeq \frac{1.506\ (2gh)^{1/2}}{a}$$

For a container of cross-sectional area $S(y)$, the equation for $y(t)$ may therefore be written, by analogy with Equation (3), as

$$- S(y)\ \frac{dy}{dt} = AC\,(2gy)^{1/2}\ (1 - e^{-\lambda t})$$

if it is assumed that the efflux velocity is precisely

$$(2gy)^{1/2}\ (1 - e^{-\lambda t})$$

This equation is still separable. Thus

$$-\int \frac{S(y)}{y^{1/2}}dy = AC\,(2g)^{1/2} \int (1 - e^{-\lambda t})\ dt$$

and hence

$$-\int_{h}^{y} \frac{S(y)}{y^{1/2}}\ dy = AC\,(2g)^{1/2} \left\{ t + \frac{1}{\lambda}\,(e^{-\lambda t} - 1) \right\}$$

We see that the maximum amount by which the predicted emptying time is changed is

$$\lambda^{-1} = \frac{0.664\ a}{(2gh)^{1/2}}$$

For the problem considered earlier, the area of the hole was 0.258 cm² so we write

$$\pi a^2 = 0.258$$

and hence

$$a = \frac{0.508}{\sqrt{\pi}} = 0.287\ \text{cm}$$

Since $(2gh)^{1/2}$ = 157 cm/s, it follows that

$$\lambda^{-1} = 0.0012 \text{ s}$$

which is insignificant.   Clearly this effect can become significant only when the relative hole size is much greater than in this problem.

4.6.   The effects of optimum tilting of the container

A second effect ignored so far is that the sides of the sachet, adjacent to the hole through which the milk poured out, were not symmetrically placed relative to the vertical.   However, although the angle of tilt was initially limited by the need to avoid milk leaking through the upper (air) hole, it gradually becomes possible, as milk pours out, to increase the angle of tilt until a position of symmetry is reached, and this was done quite subconsciously by the author in carrying out his experiment.

Now, given a prescribed volume of fluid in a specific container, the height of the fluid is greatest when the container is symmetrically placed, and this maximisation of the height of the fluid will increase the speed of efflux, with a corresponding reduction of the emptying time.   Let us calculate how significant the effect is.

Initially the position is as shown in *Figure 4.3,* with the surface of the fluid just below the corner A.   Thus θ is increased gradually from its initial value a (= 10°) to the value for which tan θ = a/b, when the surface lies precisely along the diagonal AC;   thereafter conditions are as shown in *Figure 4.4.*

In the first stage we let S be the area of the fluid surface.   Since this surface is roughly elliptical in shape, with a major axis of length b sec θ and a minor axis of approximately constant length, we may write

$$S = S_0 \cos a \sec \theta$$

where S_0 is the initial surface area.   The total volume of fluid within the container is

$$V = S (y - b \sin \theta) + \frac{1}{3} S (b \sin \theta) \qquad (11)$$

the two terms on the right-hand side being respectively the volume of a sheared cylinder of section S and height $y - b \sin \theta$, and the volume of a sheared cone of base S and height $b \sin \theta$.   The height of the surface is

$$y = a \cos \theta \qquad (12)$$

Finally,

$$\frac{dV}{dt} = -CA(2gy)^{1/2} \tag{13}$$

To solve Equations (11) to (13) it is necessary first to eliminate two of the variables V, y and θ. It is easiest to eliminate V and y, and it is found that

$$V = S_0 \cos a \left(a - \frac{2}{3} b \tan \theta \right)$$

and hence

$$-\frac{dV}{dt} = \frac{2}{3} S_0 b \cos a \sec^2 \theta \frac{d\theta}{dt} = CA(2ga \cos \theta)^{1/2}$$

This last equation separates to yield

$$\frac{3CA(2ga)^{1/2}}{2S_0 b \cos a} t_1 = \int_{\tan a}^{a/b} (1 + \tau^2)^{1/4} \, d\tau \tag{14}$$

where $\tau = \tan \theta$ and t_1 is the time taken for the first stage. The integral is readily evaluated numerically by Simpson's rule.

In the second stage the situation is as shown in *Figure 4.4.*

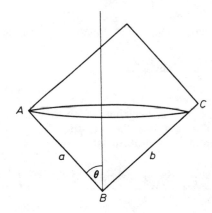

Figure 4.4   Polythene milk sachet: later stages of pouring

Initially (i.e. when $t = t_1$)

$$S = S_1 = S_0 \cos a \sec \left\{ \tan^{-1} \left(\frac{a}{b} \right) \right\}$$

$$= S_0 \cos a \left(1 + \frac{a^2}{b^2} \right)^{1/2}$$

Thereafter

$$S = S_1 \cosec^2 \theta \sin^2 \left\{ \tan^{-1}\left(\frac{a}{b}\right) \right\}$$

$$= S_1 \frac{a^2}{a^2 + b^2} \cosec^2 \theta \qquad (15)$$

and hence

$$V = \frac{1}{3} Sy$$

$$= \frac{1}{3} S_1 \frac{a^3}{a^2 + b^2} \cosec^2 \theta \cos \theta$$

since the volume is that of a sheared cone of base S and height y. We therefore have

$$- \frac{dV}{dt} = CA (2ga \cos \theta)^{1/2} = - \frac{1}{3} S_1 \frac{a^3}{a^2 + b^2} \frac{d}{d\theta} (\cosec^2 \theta \cos \theta) \frac{d\theta}{dt}$$

$$= \frac{1}{3} S_1 \frac{a^3}{a^2 + b^2} (\cot^2 \theta + \cosec^2 \theta) \cosec \theta \frac{d\theta}{dt}$$

Again we have a separable equation, whose solution is

$$\frac{3CA(2ga)^{1/2} (a^2 + b^2)}{S_1 a^3} t_2 = \int_{\tan^{-1}(a/b)}^{\pi/4} \frac{\cot^2 \theta + \cosec^2 \theta}{\sin \theta (\cos \theta)^{1/2}} d\theta$$

where t_2 is the further time taken for the angle θ to increase from $\tan^{-1} (a/b)$ to $\pi/4$. As before, the change of variable $\tau = \tan \theta$ proves helpful, and leads to the result

$$\frac{3CA (2ga)^{1/2} (a^2 + b^2)^{1/2} b}{S_0 a^3 \cos a} t_2 = \int_{a/b}^{1} \frac{2 + \tau^2}{\tau^3 (1 + \tau^2)^{1/4}} d\tau \quad (16)$$

The integral is again easily evaluated by Simpson's rule.

Once θ has increased to $45°$ it is kept constant, and thereafter the fluid empties from an elliptical cone. Equation (5) shows that it empties in a further time

$$t_3 = \frac{\frac{2}{5} y_2 S_2}{AC (2gy_2)^{1/2}}$$

where y_2 is the height and S_2 the surface area when θ first equals $\pi/4$.

Thus

$$y_2 = a \cos \frac{\pi}{4} = \frac{a}{\sqrt{2}}$$

and S_2 is given from Equation (15) as

$$S_2 = S_1 \frac{a^2}{a^2 + b^2} \csc^2 \frac{\pi}{4}$$

$$= \frac{2S_0 a^2 \cos a}{b (a^2 + b^2)^{1/2}}$$

Thus

$$\frac{CA (2ga)^{1/2} (a^2 + b^2)^{1/2} b}{S_0 a^3 \cos a} t_3 = \frac{(2)^{7/4}}{5} \qquad (17)$$

The total emptying time T then follows from Equations (14), (16) and (17) as

$$\frac{CA (2ga)^{1/2}}{S_0 a \cos a} T = \frac{2b}{3a} \int_{\tan a}^{a/b} (1 + \tau^2)^{1/4} \, d\tau +$$

$$\frac{a^2}{b (a^2 + b^2)^{1/2}} \left\{ \frac{1}{3} \int_{a/b}^{1} \frac{2 + \tau^2}{\tau^3 (1 + \tau^2)^{1/4}} \, d\tau + \right.$$

$$\left. \frac{1}{5} (2)^{7/4} \right\} \qquad (18)$$

For this problem we note or calculate in turn that

$$b/a = 1.2$$

$$(2ga)^{1/2} = 158 \text{ cm/s}$$

$$CA = 0.192 \text{ cm}^2 \qquad (19)$$

$$a \cos a = 12.51 \text{ cm}$$

$$S_0 = 52.86 \text{ cm}^2$$

and hence that

$$T = 17.43 \int_{0.1763}^{0.8355} (1 + \tau^2)^{1/4} \, d\tau +$$

$$3.905 \int_{0.8355}^{1} \frac{2 + \tau^2}{\tau^3 (1 + \tau^2)^{1/4}} \, d\tau + 7.82$$

Finally, evaluating the integrals by Simpson's rule, we find that

$$T = 12.2 + 2.1 + 7.8 = 22.1 \text{ s}$$

We see that, theoretically, we can reduce the emptying time from 28 s to 22 s, provided the container is tilted exactly in the optimum manner. In practice one could hardly achieve the whole of this 6 s improvement, so the experimentally observed 25 s is very much in line with theory.

4.7.   Time taken to tilt container

The observant reader may have noticed one further unrealistic limitation of the analysis, which assumes that the container is tilted to the position shown in *Figure 4.3* before a hole is cut at *B* through which the milk pours. In reality most normal people cut the holes first, with *AB* horizontal, and then tilt the container as fast as is practicable, until the danger of spillage at *A* limits the angle of tilt. Since a little milk is poured during this period, it will be possible to increase the angle θ in *Figure 4.3* to a value β ($>a$) before the procedure of Stage 1 is followed. It follows that Equation (18) may be re-written as

$$\frac{CA(2ga)^{1/2}}{S_0\, a\, \cos a}(T - t_0) = \frac{2b}{3a}\int_{\tan \beta}^{a/b} (1 + \tau^2)^{1/4}\, d\tau +$$

$$\frac{a^2}{b\,(a^2 + b^2)^{1/2}} \left\{ \frac{1}{3}\int_{a/b}^{1} \frac{2 + \tau^2}{\tau^3\,(1 + \tau^2)^{1/4}}\, d\tau + \frac{1}{5}\, 2^{7/4}\right\}$$

provided $\tan \beta \leqslant a/b$, where t_0 is the time taken over this initial tilting.

For our problem, with the parameters of Equation (19), this yields

$$T = t_0 + 17.43\int_{\tan \beta}^{0.8355} (1 + \tau^2)^{1/4}\, dt +$$

$$3.905 \int_{0.8355}^{1} \frac{2 + \tau^2}{\tau^3\,(1 + \tau^2)^{1/4}}\, d\tau + 7.82$$

$$= t_0 - 17.43\int_{\tan a}^{\tan \beta} (1 + \tau^2)^{1/4}\, d\tau + 22.1 \text{ s} \qquad (20)$$

The second term on the right-hand side represents time saved in the old Stage 1, since some milk has already poured out before this stage is reached.

We measure time from the instant when milk first pours from the container, which happens when the container is sufficiently tilted. Assume, for simplicity, that thereafter the height y of the fluid above the hole at B increases linearly with time, until a height y_0 (say) is reached after a time t_0.   Thus

$$y = y_0 \, t/t_0$$

Then, by Equation (13), the volume of fluid which flows from the container during this process is

$$CA \, (2g)^{1/2} \int_0^{t_0} y^{1/2} \, \mathrm{d}t = \frac{2}{3} \, CA \, (2gy_0)^{1/2} \, t_0$$

and hence, if $\theta = \beta$ say at this stage,

$$S_0 \, \cos a \, \left(a - \frac{2}{3} \, b \, \tan \beta\right) = V_0 - \frac{2}{3} \, CA \, (2gy_0)^{1/2} \, t_0 \tag{21}$$

where $V_0 = 1$ pint (0.568 litres).   We note also, from Equation (12), that

$$y_0 = a \, \cos \beta \tag{22}$$

Upon eliminating y_0 between Equations (21) and (22), and after substituting the appropriate parameters from Equation (19), it follows that

$$\tan \beta - \tan a = 0.0383 \, (\cos \beta)^{1/2} \, t_0 \tag{23}$$

For any assumed value of β, the corresponding value of t_0 follows immediately, and the value of

$$17.43 \int_{\tan a}^{\tan \beta} (1 + \tau^2)^{1/4} \, \mathrm{d}\tau$$

is evaluated numerically.   Examination of the relevant numerical values indicates that

$$17.43 \int_{\tan a}^{\tan \beta} (1 + \tau^2)^{1/4} \, \mathrm{d}\tau \simeq 0.666 \, t_0 - 0.002 \, t_0^2 \tag{24}$$

to two significant figures at least, for values of t_0 as high as 20 s.

As a specific numerical possibility, suppose it takes 2 s to achieve the initial tilting of the container.   (If the container is tilted too rapidly, the milk is liable to come out in an unexpected direction, and it spills on to the floor.   Likewise, excessive speed can lead to an over-tilting, so that milk escapes through the air hole at A; 2 s is a figure derived from the author's experiments.)   Then the time taken over the whole of the pouring operation becomes 22.8 s.

4.8.   References

BIRKHOFF, G. and ZARANTONELLO, E.H. (1957).   *Jets, Wakes and Cavities,* Academic Press; London

CURLE, N. (1956).   'Unsteady Two-Dimensional Flows with Free Boundaries' (I and II), *Proc. Roy. Soc.,* **A 235,** 375, 382

GARABEDIAN, P.R. (1956).   'Calculation of Axially Symmetric Cavities and Wakes', *Pacific J. Math.,* **6,** 611

MILNE-THOMSON, L.M. (1968).   *Theoretical Hydrodynamics,* 5th edn. Macmillan; London

4.9.   Problems for further study

1.   Use the result of Equation (5) to show that, when fluid empties from a conical container, the mean speed of efflux is five-sixths of the initial value.

2.   Likewise, for a container whose cross-sectional area at a height y is

$$S(y) = S_0 \, (y/h)^{2n}$$

show that the ratio of the mean efflux speed to its maximum (initial) value is

$$(2n + 1/2)/(2n + 1)$$

3.   Show that for sufficiently small values of t_0, when $\beta - a$ is also therefore small,

$$\int_{\tan a}^{\tan \beta} (1 + \tau^2)^{1/4} \, d\tau = (\tan \beta - \tan a)(\sec a)^{1/2} + \ldots$$

Deduce that

$$17.43 \int_{\tan a}^{\tan \beta} (1 + \tau^2)^{1/4} d\tau \approx 0.6676 \, t_0 + \ldots$$

4.   More generally, assuming that the height y of the fluid above the hole at B varies with t according to the formula

$$y = y_0 \, f\!\left(\frac{t}{t_0}\right)$$

with $f(1) = 1$, and using the parameters of Equation (19), show that

$$\tan \beta - \tan a = 0.05745 \, (\cos \beta)^{1/2} \, t_0 \int_0^1 \{f(\eta)\}^{1/2} \, d\eta$$

Making the same approximations as in Question 3, further deduce that for small enough values of t_0

$$17.43 \int_{\tan a}^{\tan \beta} (1 + \tau^2)^{1/4} \, d\tau = 1.00 \int_0^1 \{f(\eta)\}^{1/2} \, d\eta \, t_0 + \ldots$$

Suggest a simple polynomial form for $f(\eta)$, satisfying such conditions as $f(0) = 0$, $f(1) = 1$, and the conditions for a 'smooth' start and finish to the initial tilting, namely $f'(0) = 0$, $f'(1) = 0$. Verify that

$$\int_0^1 \{f(\eta)\}^{1/2} \, d\eta = 0.646$$

in this case, as compared with the value 2/3 obtained when $f(\eta) = \eta$.

5. If the pourer seeks to avoid losing milk through the air hole at A, he is likely to hold the container so that the surface is not too close to A. He will therefore, in effect, behave as though the dimension a of the container is less than 12.7 cm, and the initial angle of tilt, a, will accordingly be less than 10°. Attempt to carry through the analysis for a smaller value of a, say $a = 12.0$ cm.

5
MOLECULAR MODELS

G.G. Hall
Department of Mathematics, University of Nottingham

[Prerequisites: elementary matrix algebra]

5.1. Introduction

Organic chemistry nowadays is concerned with the diverse properties
of an enormous variety of molecules. No simple theory can be found
which includes all molecules. By selecting molecules which are closely
related it becomes much easier to develop models which illuminate the
properties of these molecules by relating them quantitatively with
geometrical or topological characteristics of the molecular graphs.

The models discussed here are motivated primarily by considerations
of the classical concepts of molecular bonds and valence structures.
Although a quantum mechanical theory of electronic behaviour may
help to explain the success of the models it is not needed to describe
them.

5.2. Planar hydrocarbon molecules

There is a large class of hydrocarbon molecules which are significant
in organic chemistry and yet are sufficiently simple both in geometry
and in electronic structure to be described in terms of models which
embody only their connectivity. These molecules are planar and have
skeletons consisting entirely of carbon atoms. Each carbon atom is
bonded to three nearest neighbours with 120° angles between bonds,
the bond lengths being approximately equal. Any neighbouring positions
not filled by other carbon atoms are filled by hydrogen atoms. Each
hydrogen has a single carbon neighbour. Many of the molecules

consist of hexagonal rings but zig-zag chains are also found. Since all the atoms in the skeleton are identical it is not surprising that many molecular properties should be determined solely by the topological properties of the skeleton. The skeleton can be represented by a planar graph whose vertices represent the carbon atoms and whose sides indicate C–C bonds. The C–H bonds are ignored. Some examples are shown in *Figure 5.1*.

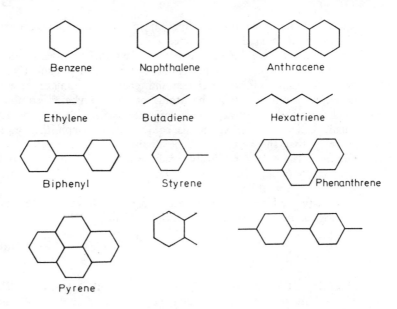

Figure 5.1   Graphs of some hydrocarbon molecules

Since the graphs convey more of the essential features of the molecule than the chemical formulae their first use is to describe the molecules.

5.3.   Graph and matrix models

The molecular graphs just described enable certain features of the molecules to be distinguished and examined. However, they are not well suited to numerical work and other models are also needed. One of these is the representation by a topological matrix T which enables the graph to be described in numerical terms. The matrix element T_{rs} is 1 when the atoms r and s are bonded and is 0 otherwise. Thus the benzene molecule, whose graph is the simple hexagon, has the topological matrix

$$T = \begin{bmatrix} 0 & 1 & 0 & 0 & 0 & 1 \\ 1 & 0 & 1 & 0 & 0 & 0 \\ 0 & 1 & 0 & 1 & 0 & 0 \\ 0 & 0 & 1 & 0 & 1 & 0 \\ 0 & 0 & 0 & 1 & 0 & 1 \\ 1 & 0 & 0 & 0 & 1 & 0 \end{bmatrix}$$

Since T contains all the information in the graph, any property of the graph can be related to a property of T.

For most of the molecules in this class the topological matrix is unnecessarily large. As long as the bond angles remain $120°$ these molecules cannot form rings containing an odd number of vertices. It follows from this that the atoms can be numbered so that any two-bonded neighbours always have opposite parity. This property is usually called *alternation* and is equivalent, in graph theory, to the property that the vertices can be distinctly coloured with just two colours. There is a relatively small number of non-alternant hydrocarbons containing five or seven numbered rings and these will not be included in the theory.

By listing all the odd numbered atoms, followed by the evens, the topological matrix assumes partitioned form

$$T = \begin{bmatrix} 0 & B \\ B^T & 0 \end{bmatrix}.$$

where B is the *adjacency matrix* and B^T is its transpose. This matrix B, contains all the non-vanishing elements of T and so all the essential information. For benzene the matrix model is now reduced to

$$B = \begin{array}{c} \\ 1 \\ 3 \\ 5 \end{array} \begin{matrix} 2 & 4 & 6 \\ \begin{bmatrix} 1 & 0 & 1 \\ 1 & 1 & 0 \\ 0 & 1 & 1 \end{bmatrix} \end{matrix}$$

Some elementary properties of the molecule can be deduced trivially from B. The total number of carbon atoms, n, must be the sum of the rows and columns of B. Similarly the total number of C–C links, N, must be the sum of all the elements in B. This is more conveniently expressed as

$$N = \text{Tr} (B^T B)$$

where Tr denotes the trace operation. A feature of the graph is the number of rings, r, which is not so obviously obtainable from B. However, r is related to n and N by the relation

$$r = N - n + 1$$

so that r can be found immediately (*see* Problem 2).

Within the constraint of alternation many ways of numbering the atoms in a molecule can be found. These will change some of the properties of B. It is important to note that some properties such as r, N, n are invariant to any change of numbering.

5.4. Free radicals

One very obvious feature of B is whether it is square or rectangular. For B to be rectangular the number of odd atoms cannot be equal to the number of even atoms. This is clear from the graph when the total number of atoms is odd, but, if n is even, it is not an obvious property of the graph. A method of recognising this situation from the graph is to note that every vertical line in the graph must join an odd and an even atom. Once the vertical lines are excluded only the external vertices pointing up (Λ) or down (V) remain to be considered. These must also have opposite parity so that the excess of one parity over the other is given by

$$|\Lambda - V|$$

and this gives the difference between the dimensions of B.

Figure 5.2 gives several examples of graphs whose B matrices are rectangular. This distinguishing of rectangular from square matrices is found to be of chemical significance. Species with rectangular B are found to be free radicals with a short lifetime because of their high reactivity and their identification from the graphs is therefore important.

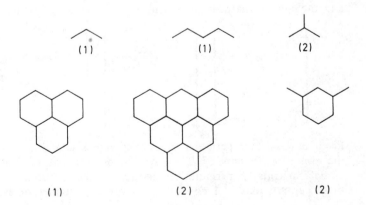

Figure 5.2   Examples of free radicals showing various parity excesses

5.5.   Structures

In the classical theory of chemical valence each carbon atom has a valence of four.   Three of these are used in forming bonds to its neighbours so that one is available to form a double bond to one of its neighbouring carbons.   The majority of these molecules have several alternative patterns of double bonds, or *structures*.   The chemical behaviour of the molecule is often explained in terms of these structures.

The relation between the structure for a molecule and its B matrix is most easily established for alternants having a square B.   To illustrate we take naphthalene

$$B = \begin{bmatrix} 1 & 0 & 1 & 0 & 0 \\ 1 & 1 & 0 & 0 & 0 \\ 0 & 1 & 1 & 0 & 1 \\ 0 & 0 & 1 & 1 & 0 \\ 0 & 0 & 0 & 1 & 1 \end{bmatrix}$$

Trial and error shows that there are three structures with the graphs

and the related matrices in which only the double bonds are noted

$$\begin{bmatrix} 1 & . & . & . & . \\ . & 1 & . & . & . \\ . & . & 1 & . & . \\ . & . & . & 1 & . \\ . & . & . & . & 1 \end{bmatrix} \begin{bmatrix} . & . & 1 & . & . \\ 1 & . & . & . & . \\ . & 1 & . & . & . \\ . & . & . & 1 & . \\ . & . & . & . & 1 \end{bmatrix} \begin{bmatrix} 1 & . & . & . & . \\ . & 1 & . & . & . \\ . & . & . & . & 1 \\ . & . & 1 & . & . \\ . & . & . & 1 & . \end{bmatrix}$$

Each of these is a permutation matrix and is constructed using only the non-zero elements of B.   Now the determinant of B is the sum of determinants of permutation matrices.   The determinant of each of these permutations is 1 so that $|B|$ gives the number of structures.

The general relation between $|B|$ and the number of structures can be established by an extension of this argument. In a structure each carbon participates in exactly one bond so that the corresponding matrix has one 1 in each row and column and constitutes a permutation matrix. Since the bonds are selected from those listed in B and since $|B|$ is a sum over all possible permutations the full set of structures must appear. The permutations involving a bond which is not a nearest neighbour bond will make no contribution to $|B|$. For rings containing $(4n + 2)$ atoms the permutations arise as cyclic permutations involving an odd number of double bonds and these have even parity. If the atoms are so numbered that one of the structures is the identity then all such determinants are +1. Thus for all such molecules the number of structures is given by $|B|$.

This simple relation breaks down if the molecule contains a ring having $(4n)$ atoms since the permutations can then be divided into equal numbers of odd and even parity. The cyclic molecule $C_{12}H_{12}$, for example, has the two structures

and these have opposite sign so that

$$|B| = 0$$

The number of structures for such a molecule is then given by the permanent

$$|B|_+$$

in which all the permutations in the product are taken with positive sign.

The subset of these molecules which have vanishing $|B|$ are distinguished by a common physical property. They are generally unstable in this geometry and will distort to a less symmetrical configuration. This distortion may involve a stabilisation of one structure by alternating long and short C–C bonds or it may involve a departure from the plane and occasionally a disintegration of the molecule. It is interesting to note that what seemed to be exclusively a mathematical peculiarity should have physical importance. The fact that this model provides such a simple test for geometrically unstable structures is an important confirmation of the value of the model.

5.6. Enumeration of structures

In some chemical discussions of these molecules it is important to know the total number of possible structures. At first this was done by drawing them but this rapidly becomes tedious for larger molecules and has the disadvantage that the set drawn cannot be proved to be complete. A direct numerical evaluation of $|B|$ is possible but gives little insight into how the different numbers vary. Other graphical methods of enumeration have been developed which are both efficient and significant. One key concept in these techniques is that, if any bond in a molecule is selected, the structures divide into two classes according to whether that bond is single or double. This fixing of the character of one bond usually forces the fixing of others. The number of structures is the sum of those in both classes and these may be easier to enumerate. This division into classes can be repeated as many times as necessary to facilitate the counting.

The polyacenes are the molecules which consist of a row of hexagons joined by edges. Benzene and naphthalene are the first two members of the sequence. In these molecules, if one vertical bond is selected and made double it is readily seen that all the remaining double bonds are uniquely fixed. The number of structures is then equal to the number of verticals and this is just $(r + 1)$, where r is the number of rings.

The polyphenyls also consist of hexagons but these are joined by opposite vertices. Two members of the series are biphenyl and triphenyl

Biphenyl   Triphenyl

It is quickly seen that there can be no structures in which a con-necting bond between rings is double. The structures in each ring can then be combined in all possible ways. Since each ring has 2 structures the total number of structures is 2^r. Thus two very similar series show a very different dependence of the number of structures on the number of rings.

Another molecular series in which the number of structures can be calculated systematically has its hexagons joined together in a staggered line as in phenanthrene and chrysene

Phenanthrene   Chrysene

The hexagon on the right is singled out and structures considered in which one of its horizontal bonds is single or double, e.g.

Double: 2 structures   Single: 3 structures

It is clear that if this bond is double then the number of structures is the same as if two rings are removed from the molecule. If it is single the number is as if one ring is removed. By repeating the argument for these smaller molecules the total can be obtained. In practice it is easier to reverse the sequence and build up the molecule from a single ring on the left. Successive rings then are added and generate a Fibonacci sequence for the number of structures (such sequences often arise elsewhere in nature, e.g. the spiralling pattern of flower petals)

5.7. Resonance energy

In classical valence theory the valencies of all the atoms are satisfied in each of the structures. The increased stability of a molecule which has several structures is said to be due to some 'resonance' between the structures. Careful experimental measurements of the heats of formation can detect this effect and experimental 'resonance energies' have been found for many of these molecules. The experiments and their analysis are difficult, however, and it is much easier for our purposes to replace them by 'resonance energies' calculated using a simple quantum mechanical theory. According to this theory the resonance energy (in appropriate units) is defined in terms of B as

$$R = 2\mathrm{Tr}\,[(B^T B)^{1/2} - I]$$

where the particular square root of the matrix is that which maximises R. Values of R have been calculated on computers for a large number of molecules. A compilation of these is given by Coulson and Streit-wieser (1965). A short table is given here to illustrate the results. We will try to model R by relating it empirically to the invariants of the graphs which we have already mentioned.

Table 5.1   Resonance energies

benzene	2	ethylene	0	biphenyl	4.383
naphthalene	3.683	butadiene	0.472	triphenyl	6.772
anthracene	5.314	hexatriene	0.988	styrene	2.424
phenanthrene	5.449	chrysene	7.190	pyrene	6.505

We consider first the molecules which consist of hexagons and have no free ends. An inspection of some values of R reveals that R is approximately proportional to N and, by using the fact that $R = 2$ for benzene, the relation

$$R \sim \frac{1}{3} N$$

is suggested. The accuracy of this simple relation is shown in *Figure 5.3*. A slightly better fit can be obtained by including a term proportional to the number of rings in the molecule. The relation

$$R \sim 0.328\ N + 0.091\ (r - 1)$$

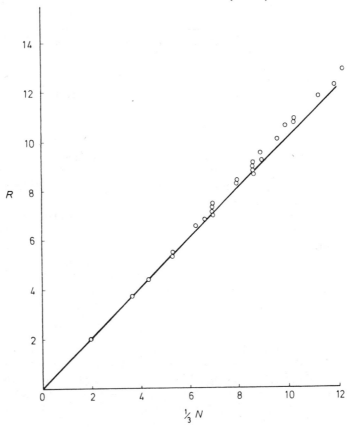

Figure 5.3   Resonance energies as $\frac{1}{3} N$

is obtained by a least squares fitting to a large number of values of
R.

The resonance energies for the remaining stable molecules are more
difficult to analyse. Since ethylene with $N = 1$ has no resonance
energy another effect must be included. We shall assume that end
effects are important and include a term proportional to the number
of ends. An approximate relation which reproduces the main trends
is

$$R \sim \frac{1}{4} N - \frac{1}{8} S$$

where S is the number of carbon atoms bonded to one other carbon.
This formula is to be applied after any complete hexagons in the
molecule have been accounted for using the formula above.

A difficulty arises in defining the resonance energy of a free radical.
It will be taken here that the unit matrix I in the formula for R has
dimension equal to the smaller of the number of rows and the number
of columns in B. A similar fitting of the theoretically calculated
values of R produces the relation

$$R \sim \frac{1}{3} N + \frac{1}{10} (r - 1) - 0.17S + 0.56$$

which gives reasonable results over a wide range of radicals.

The success of these simple relations for R raises a number of
issues. One of the most obvious of these is why other, more elaborate,
properties of the graph are not involved. The original introduction of
the term 'resonance' implied that the number of structures was important. To investigate this further we consider groups of molecules
which have common values of n and N. When the resonance energies
for the molecules in one of these groups is plotted against $|B|$ the
result is a good linear fit (*Figure 5.4*). The scale of this relation is
much finer than for the dependence on N. Thus $|B|$ becomes a signi-
ficant variable in determining the fine structure of R and no other
variable appears to be necessary.

This discussion illustrates the need to interpret the results of elaborate calculations in a meaningful way which explains the major trends
in the results and relates them to features of the situation which can
be discerned readily.

5.8. Bond orders

If the individual columns of the matrix B are regarded as vectors then
it is natural to ask whether an orthonormal set of vectors can be
found which are closely related to these vectors since orthonormal
vectors usually lead to a simpler theory. One such relation requires
that the scalar products of the vectors should be symmetrical. This

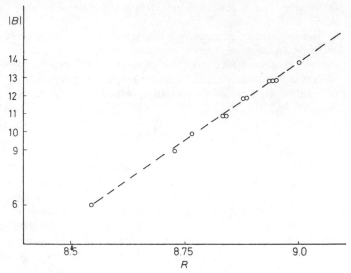

Figure 5.4   Resonance energies of 5-ring molecules

is equivalent to the relation

$$B^T P = P^T B$$

where P is the set of orthonormal vectors combined to form a matrix. The orthonormality property forces P to be an orthogonal matrix, i.e.

$$P^T P = PP^T = I$$

These two matrix equations provide sufficient relations to determine all the elements of P and it is easily verified that their solution is

$$P = B (B^T B)^{-1/2}$$

This matrix P is known as the bond order matrix and is obviously an orthogonalised form of B.

The resonance energy can be defined now as

$$R = 2\text{Tr} (B^T P - I)$$

and the first term is a sum over the bond orders of all bonds. Since this definition leads to the result

$$P_{rs} = \frac{1}{2} \frac{\partial R}{\partial B_{rs}}$$

it becomes clear that P will occur naturally in any discussion of R for molecules in which the elements of B can vary. An approximation to P leads automatically to a model for R. The simplest approximation for P is

$$P \sim (B^T)^{-1}$$

since this satisfies the first equation identically though not the second. This is a poor approximation for R since it implies $R = 0$.

The calculation of B^{-1} from B is relatively easy. Each element in B^{-1} is the cofactor of the corresponding element in B^T divided by $|B|$. The magnitude of the cofactor can be found graphically in the same way as for $|B|$ since it is a determinant with the row and column removed which intersect at the matrix element.

It is, therefore, the number of structures in the molecule that remain when the two corresponding atoms are removed. The sign of this number is most easily obtained from the alternation rule wherein nearest neighbours are +, third nearest -, fifth nearest + etc. Thus each component of $(B^T)^{-1}$ can be calculated graphically and the full result can be checked by matrix multiplication if required.

By examining the values of P and $(B^T)^{-1}$ for a number of molecules it is seen that $(B^T)^{-1}$ is a reasonably good approximation for the nearest-neighbour elements of P though it underestimates them and overestimates the remaining elements. This can be corrected by taking as the approximate P a linear combination of $(B^T)^{-1}$ and B. This also satisfies the first equation but not the second. The combination

$$P \sim (B^T)^{-1} + \frac{1}{6} B$$

increases the nearest-neighbour values by 1/6. It also leads to the simple expression for R,

$$R \sim \frac{1}{3} N$$

already discussed. A better fitting of bond orders for some simple hexagonal molecules gives

$$P \sim 0.909 \, (B^T)^{-1} + 0.210 \, B$$

and this leads to the more accurate expression for R

$$R \sim 0.328 \, N + 0.091 \, (r - 1)$$

The accuracy of these different relations for P is illustrated in *Table 5.2* for the molecule chrysene.

Table 5.2 Nearest-neighbour bond orders for chrysene

	Exact P	$(B^T)^{-1}$	$(B^T)^{-1} + \frac{1}{6}B$	$0.909\ (B^T)^{-1} + 0.210\ B$
A	0.573	0.5	0.667	0.665
B	0.476	0.25	0.417	0.437
C	0.583	0.375	0.542	0.551
D	0.707	0.625	0.792	0.778
E	0.617	0.375	0.542	0.551
F	0.712	0.625	0.792	0.778
G	0.568	0.375	0.542	0.551
H	0.535	0.375	0.542	0.551
J	0.521	0.25	0.417	0.437
K	0.754	0.75	0.917	0.892
L	0.538	0.25	0.417	0.437

5.9.   Internuclear distances

When the distances between the carbon nuclei in a molecule are measured accurately using X-ray analysis they are found to vary slightly. When plotted against the corresponding C–C bond orders the distances lie near a smooth curve.   A reasonable approximation to this curve is the linear relation

$$d \;=\; 1.517 - 0.18\ p$$

where d is the distance in Å, and p is the corresponding bond order.

   Thus the model proves capable of explaining many of the observed properties of the molecules.

5.10.   Zero eigenvalues

When the molecule contains an odd number of carbon atoms not all the valence electrons will be needed to form the bonds.   The 'unpaired' electron is spread over various atoms and its distribution will be related to the ability of the free radical to form a further bond at the various atomic sites.   This distribution can be calculated from B.

   Since the molecular system is a free radical, B is rectangular.   Let us consider the case where there is one more row than column.   The location of the 'unpaired' electron is indicated by the vector V determined by

$$V^T B \;=\; 0$$

This can be solved rapidly by taking one component of V as 1 and using the equations to find the remaining components.   The resulting vector is then normalised so that

$$V^T V \;=\; 1$$

If the dimensions of B differ by more than 1 then several independent solutions of these equations are possible and other considerations are needed to complete the discussion of the problem.

To illustrate this we take the benzyl radical

whose B matrix is

$$B = \begin{bmatrix} 1 & 0 & 0 \\ 1 & 1 & 0 \\ 0 & 1 & 1 \\ 1 & 0 & 1 \end{bmatrix}$$

By taking the component of V at 5 to be 1 the remaining components are

$$V^T = (2, -1, 1, -1)$$

and this is normalised when divided by $\sqrt{7}$. Since the reactivity is probably related to the square of the component the atom 1 will be the most sensitive.

5.11. Reference

COULSON, C.A. and STREITWIESER, A. (1965). *Dictionary of π-Electron Calculations*, Pergamon Press; Oxford

5.12. Problems for further study

1. Show that several different B matrices can be obtained for each molecule by renumbering the atoms. How may these matrices be related? Which properties of B are changed by the renumbering and which are invariant?

2. Prove that the number of rings is given by $(N - n + 1)$.

3. If S is a two-fold symmetry operation for a molecule, which turns even atoms into odd ones and vice versa, and if the even atoms are numbered by applying S to the odd ones show that B is made symmetrical. Explain why this does not result in $P = B$.

4. Prove that the numbers

$$x_s = \text{Tr}\,(B^T\,B)^s \qquad s = 0,\,1,\,2,\,...$$

are independent of how the atoms are numbered. Investigate x_2 for some molecules.

5. Devise a formula for the number of structures in the series of molecules.

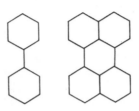

6. *The localisation energy* for a particular bond in a molecule is defined as the difference in resonance energy when that bond is removed. Show that the formulae for R provide a theory of localisation energy in terms of the nature of the remaining molecular system.

7. The behaviour of a hydrocarbon in an electrophilic reaction is often correlated with the *atom localisation energy* which is defined as the difference between R for the molecule and R for the radical in which a particular atom is removed. By evaluating this energy for some molecules devise rules for which atoms in a molecule have the largest localisation energies and which molecules contain atoms with large values.

8. Many of these molecules have the property that the matrix T has 1 as an eigenvalue. Show that this property can be interpreted graphically and hence that this eigenvalue may have high multiplicity.

6
DRILLING HOLES WITH A LASER

J.G. Andrews and D.R. Atthey
CEGB, Marchwood Engineering Laboratories, Southampton

[Prerequisites: differential equations and the Laplace transformation]

6.1. Introduction

The cutting and welding of metals is of major importance in many
areas of technology. Whilst the saw and the welding torch are still
the commonest tools for these jobs, there is considerable interest in
devising new techniques especially for special materials or where some
degree of automation is desirable. In recent years there have been
attempts to develop high-power lasers (and electron beams) for both
cutting and welding.

   The essential idea is to focus a lot of power on to a small area of
the surface of a metal, thereby producing intense surface heating and
evaporation and the subsequent formation of a hole (*see Figure 6.1*).
In cutting one tries to arrange conditions such that a hole is drilled
right through the material without molten metal slopping back into
the hole and resolidifying. In welding the situation is reversed: one
butts two pieces of metal together, heats them along their join so that
the molten metal from both sides intermixes and then resolidifies when
the heat source is taken away.

   In this chapter we consider a mathematical model of central interest
in deep penetration welding and cutting which attempts to answer the
question of how fast is it possible to dig a hole with a high-power
beam.

72

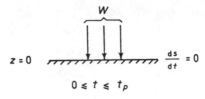

$$z = 0 \qquad \frac{ds}{dt} = 0$$

$$0 \leqslant t \leqslant t_p$$

Pre-heating of surface
to evaporation temperature

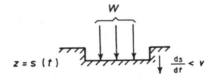

$$z = s(t) \qquad \frac{ds}{dt} < v$$

$$t_p \leqslant t \lesssim 0(t_p / \epsilon)$$

Early development of
boundary motion

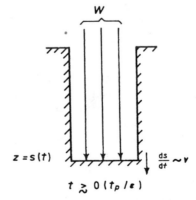

$$z = s(t) \qquad \frac{ds}{dt} \sim v$$

$$t \gtrsim 0(t_p / \epsilon)$$

Established motion of
boundary

Figure 6.1   Various stages of hole formation

6.2. Basic physical model

Consider a high-power laser beam or electron beam striking a small
area of a metal surface. Some of the energy is absorbed and the rest
is reflected. The absorption of energy takes place within a thickness
typically much less than a millimetre, so that there is surface heating
and the surface temperature rises. It does not rise indefinitely. There
are two processes which limit it. The first is heat conduction into
the material from hotter to colder metal. The second is evaporation.
When a material boils, *latent* heat is absorbed without any further rise
in temperature as the material vaporises. As the vapour puffs away
from the surface, so a hole develops in the material and it is the need
for a quantitative description of this process that calls for mathematical
modelling.

At least during the initial stages of hole formation, hydrodynamic
effects will not play a significant role. The model we consider assumes
that the process of hole formation is not complicated by such effects.
In order to show that our assumption is self-consistent, we will need
at some stage to compare the characteristic time scale for hole formation obtained from the heat conduction model with the characteristic
time scale for fluid motion. Thus the model we consider is essentially
an ablation process in which the energy of the laser which is not
reflected at the surface goes partly into evaporation and partly into
heat conduction into the material.

The simplest case to consider is that in which all the energy applied
at the surface is used to evaporate the material. This 'evaporation-
controlled limit' can arise in two ways. The first is that where the
energy is applied to the surface too rapidly for heat to be conducted
into the material. The second is that where the beam power density
is constant and the temperature distribution ahead of the evaporating
boundary approaches a steady state.

Assume that the power, W, is distributed uniformly over some area
of the surface, A, and is applied normally to the surface. ' In a time
interval δt the amount of energy dissipated is $W\delta t$ and the depth of
the hole increases by δs say. Hence the volume of material evapor-
ated is $A\delta s$ and by conservation of energy, we have

$$h\rho A\delta s = W\delta t$$

where h is the heat required to vaporise unit mass of material and ρ
is the density of the material. Rearranging and letting $\delta t \to 0$ we
obtain the speed at which the hole develops as

$$\frac{ds}{dt} = \frac{W/A}{h\rho} \tag{1}$$

Equation (1) shows that, for any given material, the limiting speed is
simply proportional to power density, W/A. For example, for a beam
power density of 1 kW/mm^2 on steel, the limiting speed is 17 mm/s.

Integrating (1) and setting $s = 0$ at $t = 0$, the depth of the hole at any instant t is

$$s(t) = \frac{1}{h\rho A} \int_0^t W \mathrm{d}t$$

or

$$s(t) = \frac{E(t)}{h\rho A}$$

where $E(t)$ is the total energy dissipated by the source in the time interval $(0,t)$. Thus, in the evaporation-controlled limit, the depth of the hole depends only on the total energy supplied to the surface.

6.3. More accurate mathematical model: allowance for heat conduction

In practice there will always be some conduction of heat into the material, so the evaporation-controlled speed (1) represents an upper limit on the rate of penetration and it is of interest to calculate the characteristic time to approach this limit. The general problem of the motion of a phase boundary with heat conduction is known as the *Stefan problem* and is well known to be difficult mathematically (Carslaw and Jaeger, 1959).

Let us suppose initially that heat is conducted in a direction normal to the surface. (The effect of sideways heat losses is left as an exercise for the student – *see* Problem (4)). Essentially we have to solve the one-dimensional unsteady heat conduction equation (Carslaw and Jaeger, 1959)

$$\frac{\partial^2 T}{\partial z^2} = \frac{1}{D} \frac{\partial T}{\partial t} \tag{2}$$

for the temperature $T(z,t)$ inside the material, where $D = K/\rho c$ is the thermal diffusivity, K, ρ and c are the thermal conductivity, density and specific heat, respectively, subject to boundary conditions at the moving boundary $z = s(t)$ and at the far face of the material.

One boundary condition at the moving boundary is obtained by applying energy conservation there, i.e.

| (rate of energy absorption by surface) | = | (rate of energy conversion into latent heat of evaporation) | + | (rate of heat conduction into the material) |

i.e.

$$\frac{W}{A} = L_v\rho \frac{ds}{dt} - K\frac{\partial T}{\partial z} \tag{3}$$

where L_v is the latent heat of evaporation per unit mass. Another boundary condition is that the temperature of the moving boundary is approximately equal to the boiling point, so that

$$T = T_v \tag{4}$$

on $z = s(t)$. In practice, for all materials, there is always a certain amount of evaporation below the actual boiling point but the vapour pressure is typically insignificant compared with atmospheric pressure except very close to the boiling point, so that (4) is a fair approximation.

For penetrating thick materials the presence of the far face is relatively unimportant and it is reasonable (and convenient) to remove the far face to infinity, where we put

$$T = 0$$

taking the ambient temperature of the material to be zero without loss of generality. For completeness it is also necessary to state boundary conditions at the other phase boundary, i.e. between the solid and the liquid. However, for many materials of interest the ratio of the latent heats of fusion and evaporation is small compared with unity (as shown in *Table 6.1*) and the discontinuity at the melting boundary can be ignored to a good approximation.

This Stefan problem has the complication (compared with the classical Stefan problem) of a power term at the moving boundary and the usual approach to solution would be by means of a numerical method (Landau, 1950; Ready, 1965). Such is the complexity of moving boundary problems with heat conduction! However, it is worth pausing a second to consider the simplicity of the solution in the evaporation-controlled limit described in Section 6.2. Can we exploit its simplicity? Intuitively, one might expect that we should be able to obtain a fairly simple solution at least in the case where the boundary is moving close to the evaporation-controlled limiting speed. In that case the characteristic heat lost by conduction to that by latent heat of evaporation

$$\left| \frac{K(\partial T/\partial t)}{L_v\rho(ds/dt)} \right| = \frac{K\ O(T_v/l)}{L_v\rho\ O(W/h\rho A)}$$

using (1), where l is some characteristic length for temperature decay into the material, given by

$$l = \frac{D}{(W/h\rho A)} \tag{5}$$

(*see* Problem (1)). Substituting for l and D, putting $h = L_v + cT_v$, we have

$$\left| \frac{K(\partial T/\partial z)}{L_v \rho(\mathrm{d}s/\mathrm{d}t)} \right| = O(\epsilon)$$

where

$$\epsilon = \frac{cT_v}{L_v} = \frac{\text{(heat required to heat material to boiling point)}}{\text{(heat required at the boiling point for evaporation)}} \quad (6)$$

is a constant for the material. Now ϵ is typically small compared with unity for many materials of interest (listed in *Table 6.1*).

Table 6.1

Material	Heat capacity (from $0°C$) cT_v (kJ/kgm)	Latent heat of fusion L_f (kJ/kgm)	Latent heat of evaporation L_v (kJ/kgm)	Ratio of latent heats L_f/L_v	ϵ cT_v/L_v
Aluminium	2 490	389	10 800	0.036	0.23
Carbon (graphite)	3 350	-	59 100	-	0.06
Copper	962	205	4 770	0.043	0.20
Gold	368	66.9	1 740	0.039	0.21
Iron	1 290	272	6 070	0.045	0.21
Lead	217	25.0	861	0.029	0.25
Magnesium	1 150	364	5 610	0.065	0.21
Mercury	41.9	11.7	305	0.038	0.14
Nickel	1 260	301	6 361	0.047	0.20
Tungsten	690	(256)	4 020	(0.063)	0.17
Water	419	335	2 260	0.15	0.19
Zinc	340	109	1 780	0.045	0.19

Consequently, the solution with $\epsilon = 0$ (i.e. where heat conduction is ignored, as discussed in Section 6.2) will be a reasonably good approximation provided the effect of heat conduction is small. For the complete solution, we look for a power series in ϵ, though in practice it is usual only to take the first two or three terms in any such perturbation expansion.

6.4. Simple perturbation solutions

We now develop a simple perturbation solution in terms of our small parameter, $\epsilon = cT_v/L_v$, to determine the motion of the boundary after long times.

In order to distinguish different orders of approximation it is convenient to introduce the following normalised variables

$$\theta = \frac{T}{T_v}, \quad \zeta = \frac{z}{l}, \quad \tau = \frac{vt}{l}, \quad \xi = \frac{s}{l} \qquad (7)$$

Taking v to be the speed of the boundary in the evaporation-controlled limit, we have from Equation (1)

$$v = \frac{W}{(cT_v + L_v)\,\rho A}$$

$$= \frac{(W/L_v\rho A)}{(1 + \epsilon)} \qquad (8)$$

The characteristic length, l, is defined by Equation (5). The normalised heat conduction Equation (2) then becomes

$$\frac{\partial^2 \theta}{\partial \zeta^2} = \frac{\partial \theta}{\partial \tau} \qquad (9)$$

with

$$\{(d\xi/d\tau) - 1\} - \epsilon\,\{(\partial\theta/\partial\zeta) + 1\} = 0 \qquad (10)$$

and

$$\theta = 1 \qquad (11)$$

on the moving boundary

$$\zeta = \xi(\tau) \qquad (12)$$

Let us try the following perturbation series for θ, ξ

$$\theta(\zeta,\tau) = \theta_0(\zeta,\tau) + \epsilon\theta_1(\zeta,\tau) + \dots \qquad (13)$$

$$\xi(\tau) = \xi_0(\tau) + \epsilon\xi_1(\tau) + \dots \qquad (14)$$

Substituting for θ and ξ in Equations (9) – (11) yields to zero order

$$\frac{\partial^2 \theta_0}{\partial \zeta^2} = \frac{\partial \theta_0}{\partial \tau}, \quad \frac{d\xi_0}{d\tau} = 1, \quad \theta_0 = 1 \qquad (15a,b,c)$$

Ignoring pre-heating effects while the boundary is being raised to its boiling point, the zero-order solution for ξ_0 is obtained from integration of (15b) as

$$\xi_0 = \tau \qquad (16)$$

To obtain the zero-order temperature distribution θ_0 we need to solve the heat conduction equation (15a) with the boundary moving at a known velocity $d\xi_0/d\tau = 1$ and with the temperature $\theta_0 = 1$ on the moving boundary. This problem can be solved fairly easily by taking the Laplace transform of equation (15a), yielding the temperature distribution as

$$\theta_0 = [e^{-(\zeta-\tau)}\text{erfc}\{(\zeta/2 - \tau)/\tau^{1/2}\} + \text{erfc}(\zeta/2\tau^{1/2})]/2 \tag{17}$$

(*see* Problem (2)), where $\text{erfc}(x)$ is the complementary error function

$$\text{erfc}(x) = (2/\pi^{1/2})\int_x^\infty e^{-y^2}\, dy$$

which is tabulated in Abramowitz and Stegun (1964).

The equations of order ϵ are

$$\frac{\partial^2\theta_2}{\partial\zeta^2} = \frac{\partial\theta_1}{\partial\tau}, \qquad \frac{d\xi_1}{d\tau} = 1 + \frac{\partial\theta_0}{\partial\zeta}, \qquad \theta_1 = 0 \tag{18a,b,c}$$

Differentiating θ_0 with respect to ζ in (17) and substituting in (18b) gives

$$\frac{d\xi_1}{d\tau} = \text{erfc}(\tau^{1/2}/2)/2 - (\pi\tau)^{-1/2}e^{-\tau/4} \tag{19}$$

Combining (15b) and (18b), we obtain the corrected normalised speed of the moving boundary as

$$\frac{d\xi}{d\tau} = 1 + \epsilon\{\text{erfc}(\tau^{1/2}/2)/2 - (\pi\tau)^{-1/2}e^{-\tau/4}\} + O(\epsilon^2) \tag{20}$$

A plot showing the variation of the (normalised) velocity of the evaporating boundary with (normalised) time is given in *Figure 6.2*. For times of $O(1)$ the velocity is reasonably close to its asymptotic value as given by Equation (1). However, for small times the solution breaks down because it predicts negative velocities! What has gone wrong? The main fault lies in our assumption that the solution with $\epsilon = 0$ (which ignores heat conduction) will be a good approximation for all times, which implies that the heat conduction term, $-K\partial T/\partial z$, in Equation (3) is small compared with the latent heat of evaporation term, $L_v\rho ds/dt$. In fact, since the boundary starts from rest (i.e. $ds/dt = 0$ at $t = 0$), the latent heat term cannot be large compared with the heat conduction term for sufficiently small times. Thus, our solution has a built-in contradiction for small times. If we inspect our solution, Equation (20), we note that $d\xi/d\tau \to -\epsilon(\pi\tau)^{-1/2}$ for $\tau \ll 1$.

To overcome this problem we must consider separately the solution for small times and re-scale the time variable and the position of the moving boundary. This is a reasonably straightforward problem in singular perturbation theory and is left as an exercise for the serious student! The full solution is presented by Andrews and Atthey (1975) but for those students who wish to have a crack by themselves some helpful texts are by Cole (1968) and Van Dyke (1964). The complete solution to first order in ϵ is

$$\frac{d\xi}{d\tau} = [1 + \epsilon\{\operatorname{erfc}(\tau^{1/2}/2)/2 - (\pi\tau)^{-1/2}e^{-\tau/4}\}] \times$$

$$[(2/\pi)\{1 + \epsilon/(\pi\tau)^{1/2}\}] \sin^{-1}\{(1 - \pi\epsilon^2/4\tau)^{1/2}\} \qquad (21)$$

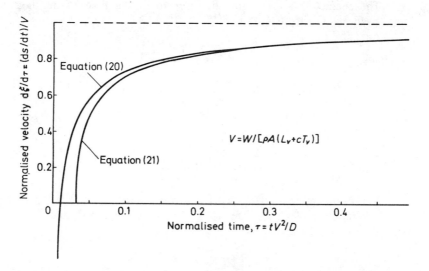

Figure 6.2 *Variation of (normalised) speed of evaporating boundary with (normalised) time for* $\epsilon = 0.2$

This is also plotted on *Figure 6.2*. Note that the solution does not begin at $\tau = 0$. This is because allowance has been made for the time required to heat the front face of the material from ambient temperature, $\theta = 0$, up to the evaporating temperature, $\theta = 1$ (*see* Problem (5a)).

6.5. Discussion

The model we have considered ignores fluid motions. Eventually, liquid on the walls will collapse into the hole and the present model will become unrealistic.

Let us estimate the effect of the gravitational force on the fluid. From dimensional arguments, the characteristic speed of the fluid moving under gravity is

$$v = \frac{dz}{dt} \sim (gz)^{1/2}$$

where z is the instantaneous depth of the hole. Integrating, putting $z = 0$ at $t = 0$, we obtain the characteristic time for fluid motion under gravity is

$$t_{\text{grav.}} \sim \left(\frac{2z}{g}\right)^{1/2}$$

The characteristic time for the laser to drill a hole of depth z is given by Equation (1) as

$$t_{\text{drill}} \sim \frac{z\rho(L_v + cT_v)}{W/A}$$

Our neglect of fluid effects is reasonable provided that

$$t_{\text{drill}} \ll t_{\text{grav.}}$$

i.e.

$$z \ll \frac{(W/A)^2}{\rho^2(L_v + cT_v)^2 g} \tag{22}$$

For example, for a laser with a power density $(W/A) = 20$ kW/mm^2 irradiating a block of steel ($\rho = 7.8 \times 10^3$ kgm/m^3 and $(L_v + cT_v) = 7.4 \times 10^3$ kJ/kgm), we require

$$z \ll 24 \text{ mm}$$

In practice it is found that the depth of penetration is limited and the above figure is in line with experimentally determined limits. Hence, it would appear that fluid effects play an important part in limiting penetration.

6.6.   References

ABRAMOWITZ, M. and STEGUN, J.A. (1956).   *Handbook of Mathematical Functions.*   National Bureau of Standards Applied Mathematics Series, 55

ANDREWS, J.G. and ATTHEY, D.R. (1975).   *J. Inst. Math. Appl.,* **15**, 59

CARSLAW, H.S. and JAEGER, J.C. (1959).   *Conduction of Heat in Solids,* 2nd edn.   Clarendon Press; Oxford

COLE, J.D. (1968).   *Perturbation Methods in Applied Mathematics,* Ginn Blaisdell; Massachusetts

LANDAU, H.G. (1950). *Q. Appl. Math.*, **8**, 81
READY, J.F. (1965). *J. Appl. Phys.*, **36**, 462
VAN DYKE, M. (1964). *Perturbation Methods in Fluid Mechanics*,
  Academic Press; New York

6.7. Problems for further study

1. Show that the characteristic length for temperature decay ahead of
a boundary at fixed temperature moving with constant velocity v into
a material of uniform diffusivity D is $l = D/v$. Hint: first show that
the steady-state heat conduction equation in moving coordinates is

$$\frac{\partial^2 T}{\partial z^2} = -\left(\frac{v}{D}\right)\frac{\partial T}{\partial z}$$

2. Derive the zero-order temperature profile given by Equation (17).

3. The problem considered in the text is that for one-dimensional
heat conduction. In practice the heat is lost in all three directions.
Under what circumstances is the one-dimensional model reasonable?

4. On a one-dimensional model any source of heat, however weak,
must produce a hole after some time interval. In practice the source
heats only a finite area and heat is conducted into the material in all
three directions. By considering a circular source of power W and
radius a heating a circular region of the surface of some material of
uniform thermal conductivity to a uniform temperature T show that
the condition for penetration is

$$W \geqslant 4aKT_v$$

(Hint: consider the analogous problem in electrostatics of a charged
conductor in a dielectric.)

5(a). Using a one-dimensional model for unsteady heat flow, show
that the time taken to pre-heat the front face of the material from
ambient temperature ($\theta = 0$) up to the boiling point ($\theta = 1$) is given
by

$$\tau = \frac{\pi\epsilon^2/4}{(1 + \epsilon)^2}$$

5(b). Solve the singular perturbation problem discussed in Section 6.4
for times $\tau = O(\epsilon^2)$ and match the small time solution with the long
time solution given by Equation (20) using Van Dyke's matching principle (Van Dyke, 1964).   (Difficult)

6.   In Section 6.5 we considered the motion of the fluid due to gravity.   What will be the effect of surface tension? Derive a condition (in terms of the surface tension coefficient) for surface tension effects to limit penetration.

7.   Consider a laser irradiating a small portion of the surface of a large pool of liquid metal.   By continuity of mass, ρv = constant across any surface.   However, since $\rho_{liquid} \gg \rho_{gas}$, the momentum, ρv^2, of the material undergoes a step change at the evaporating boundary.   In equilibrium, this change in momentum is compensated by a recoil pressure, thereby causing a depression on the surface.   Calculate the equilibrium profiles of the surface assuming that gravity is the only restoring force for (a) a source of uniform power density over a disc of radius a, and (b) a Gaussian power density distribution.   Is the neglect of surface tension reasonable?

7

STRESS ANALYSIS – STRUCTURES AND THE BIRTH OF THE FINITE-ELEMENT METHOD

O.C. Zienkiewicz
Department of Civil Engineering, University of Wales, Swansea

[Prerequisites: matrices, vector analysis and elementary mechanics]

7.1. Introduction

The problem for an engineer concerned with the design of structures
is not that of creating appropriate mathematical models but largely
lies in the solution of the models already available. The classical work
by the founders of the elasticity theory had established early in the
nineteenth century (for a survey of the foundations of elasticity, *see*
Todhunter and Pearson, 1893) the conditions of compatibility of strains
and the conditions of equilibrium of stresses in the form of differential
relations which remain unchanged today. Only the so-called 'constitutive'
relationships describing the behaviour of real materials in terms of a
stress-strain relationship leave occasional scope for new 'modelling' –
but even here the simplest of all such relations, i.e. that of linear
elasticity, has been well established. Indeed, this form of relationship
often gives an adequate basis for design on structures – but unfortunately at this point the mathematics (and mathematicians) fail the
engineer. The differential equations are capable of being solved analyti-
cally only in the most trivial cases.

Fortunately, however, amongst the 'trivial' solutions lie such problems
as those of beams and bars. Immediately, once the relations for an
individual bar have been derived, it is possible to assemble quite complex and realistic structures from such simple 'elements'. Thus, for
instance, a 'pin jointed' bar as shown in *Figure 7.1(a)* can be used as
a component of the very complex, real structure shown in *Figure
7.1(b)*. The procedures of assembly and solution of such complex
systems are purely algebraic and, although much numerical work may

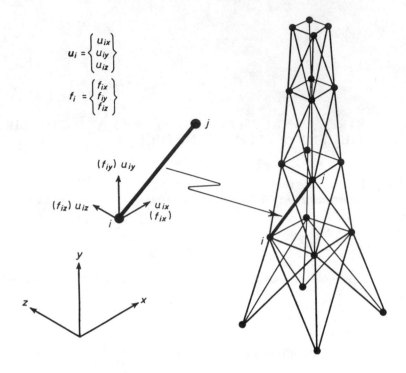

Figure 7.1   A simple bar type structure and its typical 'element'

be involved, they present little difficulty with the advent of computers. A very standardised methodology applicable to such situations has now been developed and will be briefly described in the next section.

On many occasions structures of a more 'solid' kind arise in which no simple components are available and for which we are apparently faced with a boundary value problem involving partial differential equations with a general solution not available. *Figure 7.2* shows some of such structures which may include, for example, dams, bridges, roofs and nuclear reactors.   At first sight it appears that the only way out of the dilemma is a solution for an idealised model or a crude approximation, perhaps not justified if safety and hence human lives have to be considered.   As the engineer has to 'get on with the job', irrespective of whether mathematical help is available and, as the public well knows, has to assume a responsibility for his products, some way out had to be devised.   In this the engineer reasoned that, just as complicated frame structures become amenable to analysis when sub-divided into rather basic elements, so perhaps a continuum could be similarly subdivided into small but simple 'finite elements' connected at a limited number of points (e.g. as shown in *Figure 7.2*).   His

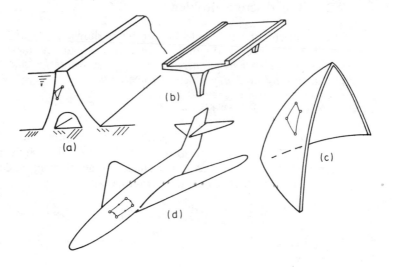

Figure 7.2   Some typical 'continued' structures

intuition led him to reason that if the basic elements were sufficiently small then sufficient accuracy would be developed in his approximation. The (intuitive) way in which properties of such finite elements were established was later 'systematised' and, indeed, it was discovered at first that the process is generally applicable to practically all mathematically formulated problems.   Today the *method of finite elements* is the subject of study not only by engineers but also by mathematicians, who are concerned with its characteristics in the 'purer sense' as a general process of approximate solution of partial differential equations.

In this chapter we shall attempt to trace some of the reasoning involved in this story and to show how the generalisation occurred. Perhaps this case of the need generating a method will be instructive.

While the illustration and derivation will be based on one example of a two-dimensional problem the reader can probably develop some generalisation for himself.   Turner *et al.* (1956) is possibly the first recorded work on the finite-element concept while the work of Zienkiewicz (1971) will give the reader some more detailed insight into the formulation and into application.   The literature on the subject today is large (with over 1000 references, the majority in the last decade) so only a basic selection is given.   References to Fraeijs de Venbeke (1965), Zlámal (1968), Oden (1969) and Strang and Fix (1973) are intended for a reader concerned with the mathematically purer aspects of the method.

7.2.   The discrete problem

The 'solution' of the simple bar problem shown in *Figure 7.1* (the element e connected at points l, m) for the case of linear elastic properties results in a relationship between the forces f_i^e and f_j^e and displacements u_i^e and u_j^e which is linear, viz.

$$f_i^e \; = \; K_{ij}^e \; u_j^e + f_{0i}^e \tag{1}$$

where u_j^e and f_i^e are vector quantities having three components in direction of appropriate axes.   Clearly when the whole structure is assembled we identify the displacement of an element node with that of the appropriate node in the structure as

$$u_i^e \; = \; u_i \tag{2a}$$

To satisfy equilibrium at any node i we require zero net force at that node, i.e.

$$\sum_{e=1}^{n} f_i^e \; = \; 0 \tag{2b}$$

where n is the total number of elements intersecting the node.   A set of linear equations is then achieved for the whole structure of the form

$$K\,u + f_0 \; = \; 0 \tag{3}$$

in which

$$u^T \; = \; (u_1{}^T, u_2{}^T, \; \dots \;)^T$$

$$f_0{}^T \; = \; (f_{01}, f_{02}, \; \dots \;)^T$$

and the reader can verify that a very simple structure of the equations occurs (*see* Problem (1)).   Now, simply,

$$K_{ij} \; = \; \sum K_{ij}^e$$
$$f_{0i} \; = \; \sum f_{0i}^e \tag{4}$$

summations being taken over all elements.

The complete solution of the assembly thus presents no more difficulty than the solution of a large, banded symmetric equation system – a matter to which engineers have put their minds.   With present day computers such a system of equations with a thousand unknowns u can easily be solved and indeed 10-20 000 unknowns can be dealt with on large computers.

7.3. The plane stress analysis problem and the first finite element

The 'finite-element' discretisation of a continuum will be illustrated in a 'plane' problem in which the stresses and displacements vary in only two dimensions. Such a problem may, for instance, be realistic in the stress analysis of the long dam illustrated in *Figure 7.2(a)* where one section must behave identically with any other.

In two dimensions we are concerned with a symmetric stress system $\boldsymbol{\sigma}$ defined by three components (*see Figure 7.3(a)*)

$$\boldsymbol{\sigma}^T = (\sigma_{xx}, \sigma_{yy}, \sigma_{xy})^T \tag{5}$$

If the body forces are expressed by a vector

$$b^T = (X, Y) \tag{6}$$

with X and Y being forces per unit volume in x, y directions, we find that in order that equilibrium is satisfied the following equation has to be satisfied in the problem domain Ω:

$$L\,\boldsymbol{\sigma} + b = 0 \tag{7}$$

in which L is a differential operator explicitly given as

$$L = \begin{bmatrix} \dfrac{\partial}{\partial x}, & 0, & \dfrac{\partial}{\partial y} \\[2ex] 0, & \dfrac{\partial}{\partial y}, & \dfrac{\partial}{\partial x} \end{bmatrix} \tag{8}$$

and $L\boldsymbol{\sigma}$ represents the force acting on unit volume of material. With the displacements being specified at any point by a two-dimensional vector u

$$u^T = (u, v) \tag{9}$$

where u, v are the components of u in the directions of the co-ordinate axes. We can represent the 'strain' or deformation occurring at any point by a vector $\boldsymbol{\epsilon}^*$

$$\boldsymbol{\epsilon}^T = (\epsilon_{xx}, \epsilon_{yy}, 2\epsilon_{xy})^T \tag{10}$$

and

$$\boldsymbol{\epsilon} = L\,u \tag{11}$$

*The vectorial representation of tensor equations, usually written as σ_{ij}, ϵ_{ij} is particularly convenient in our context – the factor 2 in Equation (10) is to account for symmetry of the strain tensor.

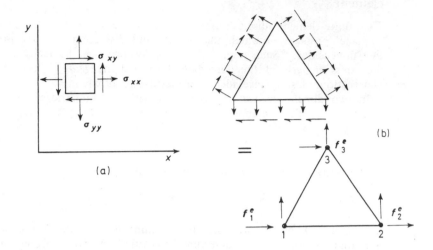

Figure 7.3 *Plane problem in elasticity (a) Definition of stress components (b) Replacement of tractions by 'nodal forces' in a finite element*

In Equation (11) the operator L is precisely the same one which occurred in the equilibrium relations. (The reasons for this need not concern us at the moment.)

The equations defining the boundary value problem are now complete and we have to consider their solutions.

Supplementing the Equations (5) to (10) with appropriate boundary conditions (which we shall not discuss here) and a constitutive relation linking σ and ϵ, properly formulates the problem which, as we have mentioned, has only been solved exactly for a very limited number of configurations. For present purpose we shall write the constitutive law for a linear elastic material as

$$\sigma = D \epsilon \qquad (12)$$

in which D is a symmetric, 3×3, matrix of elastic coefficients determined by experiment.

In attempting to formulate a physical 'finite' element the engineer isolates from a continuum a simple small shape, a triangle in the case of *Figure 7.3(b)*, and assumes that the displacements of 'nodes' 1-2-3 of the element, i.e. u_i^e, $i = 1,2,3$, will suffice to define the 'forces' at the corresponding nodal points, f_i^e, $i = 1,2,3$. Clearly these forces are only a statically equivalent set of the real tractions or stresses acting on the faces of an element.

The question of how to proceed logically at this point to relate these sets of quantities was therefore paramount. We give here one of the lines of thought that was pursued early in the history of the finite-element process.

First, we shall assume that within the element the displacements are prescribed by a set of interpolating functions. Thus for the element e shown

$$u = \begin{bmatrix} u \\ v \end{bmatrix} = \sum N_i \, u_i \qquad (13)$$

in which $N_i = I \, N_i$ and N_i are simple linear interpolations.

To obtain a statical equivalence of 'nodal forces' and internal stresses the virtual work principle is used (*see,* for example, Goldstein, 1959). Thus, if the vector δu^e represents a virtual set of nodal displacements, then by equating the external virtual work done by the nodal forces to the virtual internal work done by stresses, we have

$$\delta u^{e^T} f^e = \int_\Delta \delta \epsilon^T \, \sigma \, dx \, dy - \int_\Delta \delta u^T \, b \, dx \, dy \qquad (14)$$

the integration being then over the area of the triangle. From Equation (11), we have

and from (13)

$$\delta \epsilon = L \, \delta u$$

$$\delta u = \sum N_i \, \delta u_i \qquad (15)$$

Further, as the virtual work statement is valid for any δu_i^e we conclude that

$$f_i^e = \int_\Delta (LN_i)^T \, \sigma \, dx \, dy - \int_\Delta N_i^T \, b \, dx \, dy \qquad (16)$$

which, together with the constitutive relation (12), the displacement assumption (13) and the definition of strains (11) reduces precisely to the same form as that given by Equation (1) in the case of bars with

$$K_{ij} = \int_\Delta (LN_i)^T \, D \, (LN_j) \, dx \, dy \qquad (17a)$$

$$f_{0i} = -\int_\Delta N_i^T \, b \, dx \, dy \qquad (17b)$$

With this discretisation achieved the coefficients for each triangular element can be very simply calculated and the structure solved by precisely the same methods (and programs) as those developed for bar (discrete) systems. We shall not dwell here on the exact form of the interpolation function or the value of the element integrals. The reader can establish these for himself or consult Turner *et al.* (1956). The important points are:

1. A 'reasonable' procedure was established for deriving an analogy of discrete elements from continuum 'elements'.
2. A procedure was found which when tested against experimental or closed-form results approximated them with any degree of accuracy desired.

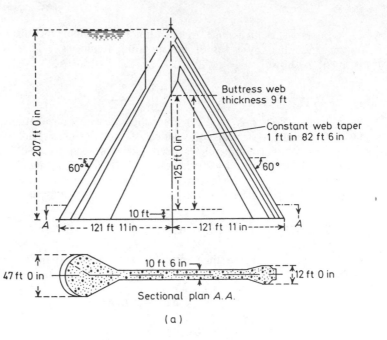

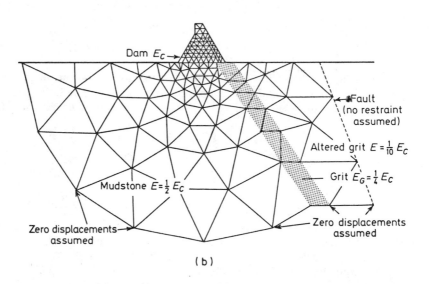

Figure 7.4 Stress analysis of a buttress dam. Plane stress condition assumed in dam and plan in foundation. (a) The buttress section analysed. (b) Extent of foundation considered and division into finite elements (from Zienkiewicz (1971); courtesy McGraw-Hill)

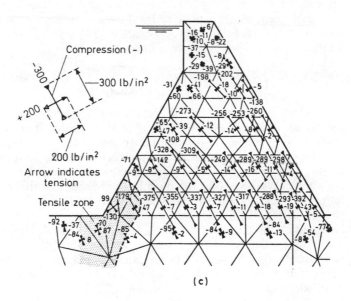

(c)

Figure 7.4 Stress analysis of a buttress dam. Plane stress condition assumed in dam and ·plan in foundation. (c) Stress analysis of the buttress dam. Principal stresses for gravity loads combined with water pressures, which are assumed to act as external loads (from Zienkiewicz (1971); courtesy McGraw-Hill)

A process was now available by which stresses could be found in real structures for which the engineer had to assume responsibility. In *Figure 7.4* we show some results of this stress analysis process applied to a dam in the design stages (during 1962-3), the structure has now been completed and is behaving satisfactorily.

7.4.   Some obvious questions and the generalisations

While the 'empirically' minded may at this stage be reasonably satisfied with the results a nagging set of questions persists:

1. Is it 'reasonable' to replace a set of continuously distributed tractions by a set of equivalent concentrated forces?
2. Will the process always converge to the 'exact' solution as the size of the elements decreases?
3. Is there any means of estimating the approximation error?
4. What is the mathematical basis for the procedure?

The answer to the final question is perhaps most revealing.   Let us assume that Equation (13) is valid for the *whole* domain Ω with the

index i now taking up the numbers of any point within it. N_i becomes
simply a *piecewise* defined function within the domain and if the
integrals of Equation (14) are taken over the whole domain rather than
over each element the inter-element forces f_i do not appear in the problem and the left-hand side of Equation (14) refers only to boundary
traction work.   Further if we assume that

$$\int_\Omega (\quad)dxdy = \sum \int_\Delta (\quad)dxdy \qquad (18)$$

for the integrands occurring in Equation (14) (summation over all elements) we have obtained the assembly rule of the discrete system
implied in Equation (2b) without mentioning the inter-element force.
Not only have we eliminated the physically dubious notion but have
at the same time imposed a condition on the interpolation functions
N_i which have been used to describe the displacements within each
'element'.   Clearly for the integration rule (18) to hold these functions,
though defined in a piecewise manner, must possess certain inter-element
continuity.   In the case discussed, with only first derivatives of N_i
involved, continuity of the function has to exist (C^0 continuity).   It
is fortunate that in using a linear interpolation in the triangular element and connecting nodal values of the function just this kind of
continuity was imposed – perhaps originally by accident – in the element
we have discussed.

More elaborate arguments show that the enforcement of this continuity, while at the same time requiring certain 'completeness' criteria on
the approximation of Equation (13), ensures indeed that convergence
occurs to the exact solution – and an answer to the second question
is provided.

At this stage the reader may observe that any problem governed
by some (partial) differential equations can be stated in an integral
form as

$$F(\phi) = 0 \rightarrow \int_\Omega v \; F(\phi) \; d\Omega = 0 \qquad (19)$$

in which ϕ is the unknown function and v any suitably continuous function, then the line of thought developed for the elastic stress analysis
problem is capable of being applied to its approximate solution.   For
if we expand ϕ in a piecewise manner in terms of some unknown
parameters a_i, i.e.

$$\phi = \sum N_i \; a_i \qquad (20)$$

then Equation (19) will yield immediately a suitable set of approximations if a system of linearly independent functions v_j is used to obtain
a set of new algebraic equations equal to the number of unknown parameters.   The integral relationships are often provided by an *extremum
principle* associated with the governing differential equations.

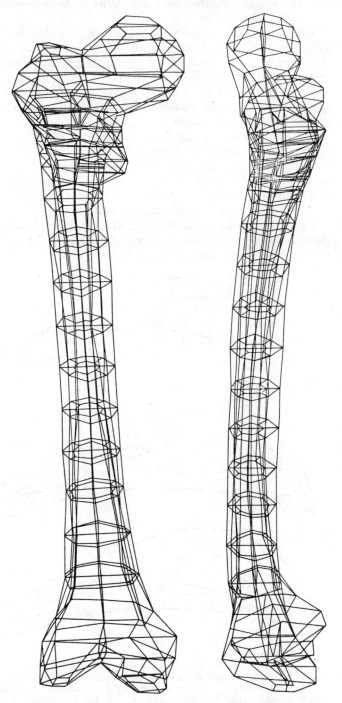

*Figure 7.5   A problem of bio-mechanics.   Plot of linear element form only —
curvature of elements omitted.   Note degenerate element shapes*

Returning to the elasticity problem, we find that the virtual work statement is in fact identical to minimisation of *total potential energy*. And here at least a part of the answer is provided for the third question of estimating errors approximately. We shall not go into this, but refer the reader to the early work of Synge (1957) who in a limited sense from the mathematical viewpoint anticipated the possibility of the finite-element method.

The generalisation of the finite-element process, which is described above, clearly has opened doors not only to the solution of the structural problems but also to a host of applications for which an integral statement can be made. The possible variants are enormous. Different approximating functions and shapes of element subdomains can be used and different physical problems dealt with.

Figure 7.5 shows for instance curvilinear 'blocks' used in three-dimensional stress analysis solutions while *Figure 7.6* shows an application to a quite complex problem of heat transfer.

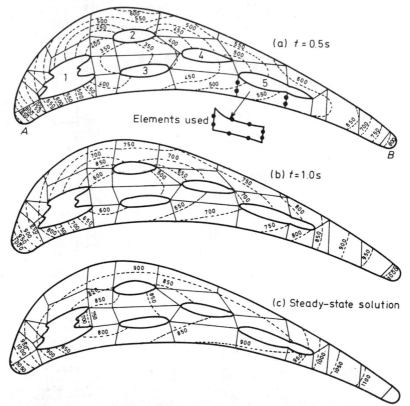

Figure 7.6   Temperature distribution in a cooled rotor blade, initially at zero temperature (Δt = 0.01 s).   Specific heat c = 0.11 cal/cm. °C; density ρ = 7.99 gm/cm³; conductivity k = 0.05 cal/s cm °C; gas temperature around blade = 1145 °C.   Heat transfer coefficient α varies from 0.390 to 0.056 on the outside surfaces of the blade (A-B)

Hole number	Cooling hole temperature	α around perimeter of each hole
1	545 °C	0.0980
2	587 °C	0.0871

(from Zienkiewicz (1971); courtesy McGraw-Hill)

7.5. Concluding remarks

This rather incomplete story of the development of the finite-element method is one in which the engineering need led to a discovery (and sometimes re-discovery*) of structural models of general applicability and for which initially at least much of the mathematics had to be developed by the engineer himself. The process of first *doing* and then *justifying* later is one in which progress is sometimes more rapid and which often can lead to unexpected results. While condensed here to a few brief statements, the establishment of 'sufficient' conditions of convergence was a matter — perhaps fortunately — of some years. In that period these sufficient conditions were repeatedly violated leading to some elements which cannot be justified on the simple arguments produced here and which we *now* know as convergent and efficient. Indeed progress is still being made on completely new lines — some waiting for a theoretical basis while being widely used at the moment.

From the modelling point of view, the interest lies not so much in the physical representation (which is essentially described by the stress-strain relationship) as in the technique employed for solving the physical equations. The art of the finite-element method comes in choosing elements which model a particular structure both accurately and economically.

7.6. References

COURANT, R. (1943). 'Variational Methods for the Solution of Problems of Equilibrium and Vibration', *Bull. Amer. Math. Soc.,* **49**, 1-23

FRAEIJS DE VEUBEKE, B.M. (1965). 'Displacement and Equilibrium Models in the Finite Element Method' in *Stress Analysis.* Eds. O.C. Zienkiewicz and G.S. Holister. Wiley; New York

GOLDSTEIN, H. (1959). *Classical Mechanics.* Addison Wesley; Reading, Massachusetts

ODEN, J.T. (1969). 'A general theory of finite elements', *Int. J. Num. Meth. Engineering,* **1**, 247-259

STRANG, G. and FIX, G. (1973). *An Analysis of the Finite Element Method.* Prentice Hall; Englewood Cliffs, New Jersey

SYNGE, J.L. (1957). *The Hypercircle in Mathematical Physics,* Cambridge University Press; London

TODHUNTER, I. and PEARSON, K. (1973). *History of the Theory of Elasticity and Strength of Materials.* Dover; New York

TURNER, M.J., CLOUGH, R.W., MARTIN, H.C. and TOPP, L.J. (1956). 'Stiffness and Deflection Analysis of Complex Structures', *J. Aero. Sci.,* **23**, 805-823

*The first recorded use of piecewise defined approximation with triangular elements was published by Courant (1943)

ZIENKIEWICZ, O.C. (1971). *The Finite Element Method in Engineering Science,* 2nd edn.   McGraw-Hill; New York
ZLÁMAL, M. (1968). 'On the finite element method', *Num. Math.,* **12**, 394-409

7.7.   Problems for further study

1.   Given that the elongation, Δ, of a bar is

$$\Delta = \frac{LP}{EA}$$

where L is the length
  E is the elastic modulus
  A is the cross-sectional area
  P is the axial force

(a) Establish stiffness coefficients K_{ij}^e for a bar for which the co-ordinates of the end points are known, and (b) assemble equations governing the behaviour of a structure shown below

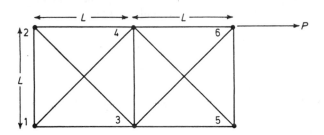

in which $u_1 = v_1 = 0$, $u_5 = v_5 = 0$ and a horizontal load P is acting at point (6).

2.   Show that the general form of Equations (1) and (4) are applicable to a network of resistances.

3.   Taking a triangular element of *Figure 7.3* establish the linear inter-polation function in detail and find the expressions (17a,b) in an explicit form.

4.   Show that a problem governed by a differential equation

$$F(\phi) = \frac{\partial}{\partial x}\left(k\,\frac{\partial \phi}{\partial x} \right) + \frac{\partial}{\partial y}\left(k\,\frac{\partial \phi}{\partial y} \right) + Q = 0$$

where $\phi = 0$ on boundaries and $k = k(x,y)$ and $Q = Q(x,y)$ can be discretised by use of Equation (19) with

$$\phi = \sum N_i \, \phi_i$$

and

$$v = N_j$$

Show that in the 'standard form'

$$K^e_{ij} = \int \left(k \, \frac{\partial N_i}{\partial x} \, \frac{\partial N_j}{\partial x} + k \, \frac{\partial N_i}{\partial y} \, \frac{\partial N_j}{\partial y} \right) \mathrm{d}x \; \mathrm{d}y$$

(Hint: Use integration by parts.)

8
POPULATION MODELS

G. Murdie
Department of Zoology and Applied Entomology, Imperial College, London

[Prerequisites: ordinary differential equations]

8.1. Introduction

The population problem has been a world concern since Malthus (1798) proposed his 'gloomy doctrine' that the human race could only persist if periods of exponential growth were punctuated by plague and famine. Much more recently, Ehrlich and Ehrlich (1970) and Meadows *et al.* (1972) have considered the exponential rise of populations and the depletion of non-renewable resources in much the same light.

Although we cannot deny the importance of the human dilemma, it is unrealistic to consider that the Malthusian model is the only one that applies to biological populations. In fact, it is a fair description for only very short periods of an organism's life cycle and has little universal value.

It is possible to develop models which represent specific growth events with a certain degree of precision. It is much more difficult to develop models with a general range of applicability. Then what use can be made of models and of what form should they take?

A whole spectrum of models can be recognised which ranges from the descriptive, empirical type to the general form (May, 1972). This distinction is the more important when applications are considered. Conway and Murdie (1972) describe models which can be applied to pest control but which range from teaching tools to those which deal with very small portions of the pest system, for example reproduction and searching behaviour, to a simulation of the field population of a single pest of cotton. These models were designed to answer specific questions and not to examine general ecological principles. In Holling's (1966; 1968) terminology these would be called 'tactical' models.

The general 'strategic' model (Holling, 1966) 'sacrifices precision in an effort to grasp at general principles' (May, 1972). Strategic models are the concern of this chapter. They provide a framework of basic biological processes, such as birth and death, in formal terms. Analysis of the model is used to gain insight into these biological processes, particularly of interactions, which might then indicate areas where research could elucidate fundamental mechanisms of control (of numbers) in natural populations.

This chapter is not intended to provide an overview of the models used in all branches of biology. The emphasis on population models does not represent an objective view of their importance relative to other biological subjects but merely the bias of an ecologist interested in the vast problem of quantifying natural population processes especially in the area of insect pest control.

The modern concern for pollution and the use of toxic chemicals to control pests has added weight to those scientists who argue for some measure of biological control by using the natural enemies of pests to full advantage. To use natural enemies we need to know something of the population dynamics of at least two-species, and most probably, multispecies systems. However, before tackling the complexities of these interactions it is advantageous to have some basic understanding of the relatively simple one-species system.

8.2.   Single-species populations

The simple Malthusian model for the rate of population growth of an organism reared in constant conditions is simply

$$\frac{\mathrm{d}N_t}{\mathrm{d}t} \;=\; bN_t \tag{1}$$

where b is the instantaneous birth rate per head of the population. Integrating, putting $N = N_0$ at $t = 0$, yields the instantaneous population size as

$$N_t \;=\; N_0 \exp(bt) \tag{2}$$

Death can be included in the model by introducing an instantaneous death rate, d, and putting $r_m = b - d$,

$$N_t \;=\; N_0 \exp(r_m t) \tag{3}$$

The parameter r_m is thus a measure of the innate capacity for population increase (Andrewartha and Birch, 1954).

Under natural conditions it is unusual to deal with periods which are short enough for r_m to be considered constant. In general, $r_m = r(t)$ and

$r_m = (b-d)$

$$N_t = N_0 \exp\left(\int_0^T r_m(t)\ dt\right) \tag{4}$$

Although unlimited geometric population increase has been suggested for natural populations (e.g. World human population: Malthus, 1798; Meadows *et al.*, 1972) we are forcibly reminded that natural resources for example, food, space and water, are limited. These constraints impose an upper limit, K say, for population size to which the population tends and on reaching or overshooting it, falls again as (for instance) food shortage leads to starvation, increased mortality and decreased fertility. The rate at which the depleted resources can be replaced, e.g. by producing more food, determines the time scale for population revival. The outcome of such processes is that populations do not maintain a steady state but fluctuate, sometimes very wildly, around a· mean level somewhat below the maximum capacity of the environment describing a series of oscillations. The study of mechanisms for stabilising these oscillations demands the development of population models.

From Equation (3) $N_t \to \infty$ as $t \to \infty$ when $r_m > 0$ and $N \to 0$ as $t \to \infty$ when $r_m < 0$; if $r_m = 0$ then the population is stable with no growth or decay. The introduction of small perturbations around the value $r_m = 0$ would make the population oscillate without control around an unstable equilibrium point. It is more usual for there to be restraints on the population growth rate which are related to the size of the population, say N_t at time t, that is to say that there is a built-in density dependent regulation. Thus the rate of change

$$\frac{dN_t}{dt} = (b - d - cN_t)N_t \tag{5}$$

where c is a constant, suggests a modification of r_m which is essentially proportional to $(K - N)/K$. A linear transformation of the kind

$$r = r_m\left(\frac{K - N_t}{K}\right) \tag{6}$$

simply means that as N approaches the carrying capacity, K, of the environment, r tends to zero, and to the maximum possible for the species, r_m, as N tends to zero; the quantity $(K - N)/K$ is a proportionate measure of the total resources unutilised. Rewriting Equation (5) as

$$\frac{dN_t}{dt} = r_m\left(1 - \frac{N_t}{K}\right)N_t \tag{7}$$

and integrating, we obtain the so-called Verhulst-Pearl logistic Equation (*Figure 8.1*)

$$N_t = \frac{N_0 K \exp(r_m t)}{K - N_0\{1 - \exp(r_m t)\}} \tag{8}$$

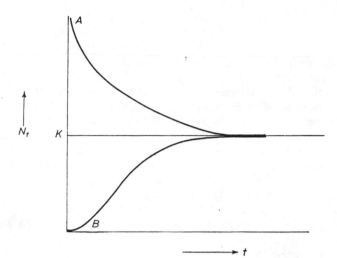

Figure 8.1   The logistic function with K *equilibrium (a)* N(0) > K, *population approaches* K *asymptotically from above* r < 0   *(b)* N(0) < K, *population approaches* K *asymptotically from below* r > 0

(Verhulst, 1838; Pearl, 1927) (*see* Problem (1)).   When

$$N_t > K, \; r \text{ is negative}$$

and when

$$N_t < K, \; r \text{ is positive}$$

and thus the population approaches K, the equilibrium density, from above or below.   At $N_t = K$, $r = 0$ and there is a stable equilibrium around which the population fluctuates; any perturbations are nullified by opposite forces proportional to $K - N_t$ bringing the population back towards the equilibrium level.   Linear corrections to r_m are of course not essential to the argument since the principle still holds for curvilinear functions.

One serious criticism of the models outlined above is the use of the instantaneous rates which suppose a perfect temporal dependence on the population at that instant.   Invariably birth is dependent on events sometime in the past and the population size at that time is probably more relevant to the regulation of the birth process.   Thus there is a time lag between the time of inception and birth.   On the other hand, it is probably less in error to consider that the death rate depends on the population conditions at any particular instant, although it is possible to imagine that some dietary deficiencies in the past could affect present susceptibility to disease and therefore mortality.   There are obvious time lags in nature, such as the extended periods of immaturity in mammals where the number becoming mature will depend on the number in the population one development period ago and thus the number born at time t is affected by conditions at $t - \tau$.   With time lags the logistic becomes a differential delay equation

$$\frac{dN_t}{dt} = b_0 \left(1 - \frac{N_{t-\tau_1}}{K} \right) N_{t - \tau_2} - d_0 N_t \qquad (9)$$

where τ_1 and τ_2 are time lags and b_0 and d_0 are base birth and death rates.

The above treatments are adequate for species with discrete generations (e.g. when breeding occurs only at certain times of the year). However, when generations overlap we must recognise that some populations will have a definite age structure, for example, juveniles (pre-reproductives) and adults which have very different mortality rates and since the juveniles do not reproduce they make very different contributions to the population. The idea of age-specific characteristics in a population has long been recognised by demographers and actuaries and has led to the development of methods peculiar to age-structured populations although there are general properties which are common to all populations.

In general we can say of all population structures that, if l_x is the probability of an individual surviving to age x, then $l_0 = 1.0$ and $l_\infty = 0$. It is convenient to consider only the female (or reproductive) offspring of the population and to define the number of female offspring born per individual adult female of age x as m_x; total population size can be obtained simply by multiplying the number of females by the sex ratio. The rate of population increase between generations, using a generation interval of 1, is

$$R = \sum_0^\infty l_x m_x = \frac{N_{t + 1}}{N_t} \qquad (10)$$

If we further define $n_{x,t}$ as the number of individuals between the ages x and $x + 1$ at time t, then the number of new-born individuals born into the population at time t is

$$n_{0,t} = \sum_{x = 0}^\infty n_{x,t} m_x \qquad (11)$$

An important question in age-structured populations is whether stable age distributions are possible. If a stable age distribution does exist there is a constant rate of increase, R, between generations, such that

$$R = \frac{\Sigma n_{x,t + 1}}{\Sigma n_{x,t}} \qquad (12)$$

then for populations T periods (e.g. generations) apart

$$n_{x,t} = n_{x,t - T} R^T \qquad (13)$$

and since

$$n_{0,t - x} = n_{0,t} R^{-x} \qquad (14)$$

and thus

$$n_{0,t} = n_{0,t} \sum R^{-x} l_x m_x \qquad (15)$$

then

$$1 = \sum R^{-x} l_x m_x \qquad (16)$$

defines R for a given set of survivorship (l_x) and age-specific birth rate (m_x) tables at stable age distribution (*see* Problem (2)). When conditions are constant and more than one age group reproduces, the population values will have damped oscillations and a stable age distribution will be approached with constant proportions of individuals in each age group.

Two additional characteristics of life-table data deserve mention:

1. The proportion of the population surviving to age x but dying in the interval x to $x + 1$ is

$$d_x = -l_{x+1} + l_x \qquad (17)$$

and then

$$l_{x+1} = 1 - \sum d_x = l_{x-1} - d_{x-1} \qquad (18)$$

2. The age-specific mortality, q_x, the probability that an individual dies in the interval x to $x + 1$, is

$$q_x = \frac{d_x}{l_x} \qquad (19)$$

When there are constant life tables, i.e. of birth and death rates in non-limiting conditions, the population will always reach the stable age distribution. When this is attained the rate of increase will approach the constant finite rate, R, which in the continually reproducing population is estimated by e^{rm}. To estimate r_m, Equation (16) can be modified to a continuous distribution, so that

$$1 = \int_0^\infty e^{-rmx} l_x m_x \, dx \qquad (20)$$

Although both R and r_m can be evaluated it must be emphasised that they apply only when the stable age distribution is achieved. However, even though the stable age condition is rarely attained, these two parameters are useful measures of the capacity of a species to increase.

8.3. Two species models

The single species models provide some insight into the mechanisms
involved in oscillations, such as the self-regulatory density dependence.
However, it is quite obvious that no organism can live in complete
isolation from others and in living together in the same habitat organisms often compete. They compete for such limited resources as food,
space, air etc., and in some cases one may kill another as a matter of
protection rather than as a specific food source. More specialised
competitions are shown by predator-prey and parasite-host systems
where one species feeds on another removing it from the habitat, and
indeed some natural enemies are so specific that they are able to feed
only on one species in order to sustain life and to reproduce. Here
we are concerned more with insect species where usually a single predator requires a number of prey to complete its life cycle, much in
the same way as a spider does, while a parasite (more correctly called
a parasitoid in this context) destroys its host completely in completing
(usually larval) development. Most insect parasites develop one offspring per host giving a convenient one-to-one correspondence between
the number of hosts attacked by the parasites and the number of
parasites realised in the next generation.

8.3.1. COMPETING SPECIES

If we consider the continuous deterministic models for two species
living separately and independently the rates of change of their populations can be written from (11) as

$$\frac{dN_1}{dt} = N_1(r_1 - b_{11} N_1)$$

$$\frac{dN_2}{dt} = N_2(r_2 - b_{22} N_2)$$

$$(21)$$

These logistic forms use the two coefficients b_{11} and b_{22} to measure
the effect of a species on its own rate of increase. When the two
species interact in some way so that their realised increase is related
not only to their own density but also to that of the other species,
then

$$\frac{dN_1}{dt} = N_1(r_1 - b_{11} N_1 - b_{12} N_2)$$

$$\frac{dN_2}{dt} = N_2(r_2 - b_{22} N_2 - b_{21} N_1)$$

$$(22)$$

where $b_{12}\,N_2$ measures the inhibiting effect of species 2 on species 1. and vice versa.

When both dN_1/dt and dN_2/dt equal zero then the two species coexist in equilibrium densities. When perturbed by a small amount the population densities either return to the equilibrium point (= stable equilibrium), or they move further away and eventually one or both of the species goes to extinction (= unstable equilibrium).

In the general time-dependent regime, the competition equations cannot be solved explicitly and it is convenient to examine the conditions graphically (*Figure 8.2*). In the absence of one or the other species two equilibria can be recognised:

$$N_1 = 0, \qquad N_2 = \frac{r_1}{b_{22}}$$
$$N_2 = 0, \qquad N_1 = \frac{r_2}{b_{11}} \tag{23}$$

A third equilibrium point exists at

$$\left.\begin{array}{l} r_1 - b_{11}\,N_1 - b_{12}\,N_2 \\ r_2 - b_{22}\,N_2 - b_{21}\,N_1 \end{array}\right\} = 0 \tag{24}$$

These models can be used to describe the interactions between any competing organisms. de Wit (1960), investigating competition between weeds, used the numbers of seeds produced at the end of the growing season as measures of population sizes.

8.3.2. HOST-PARASITE SYSTEMS

We have dealt with a simple treatment of the two-species system where one species competes with another for one or more resources which are limited in supply. The host-parasite and prey-predator systems represent more complex interactions where development of the attacking species depends absolutely, or in part, on the numbers of the food species available to it. Conversely the rate of increase of the food species depends on the number of individuals removed from its population by the parasite or predator populations. The development of parasite-host models is illuminating in that it highlights important biological and mathematical properties of the system.

If we assume that births of parasites depend on the host number, N, and that host deaths are proportional to parasite numbers, P, then the rates of change of the populations are given by

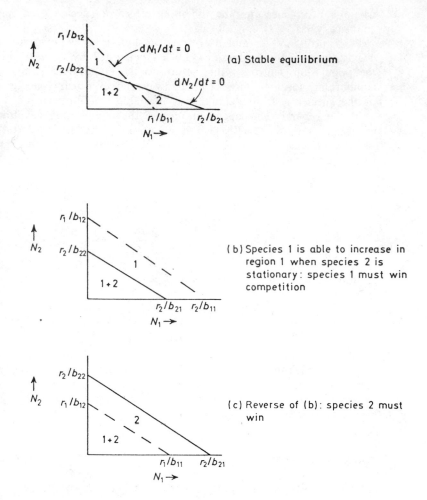

(a) Stable equilibrium

(b) Species 1 is able to increase in region 1 when species 2 is stationary: species 1 must win competition

(c) Reverse of (b): species 2 must win

(d) Unstable equilibrium

Figure 8.2 Graphical analysis of Lotka-Volterra competition equations. Numbers indicate regions in which species 1 can increase (1) and species 2 can increase (2)

$$\frac{dN_t}{dt} = (r_n - c_1 P_t) N_t$$

$$\frac{dP_t}{dt} = (- r_p + c_2 N_t) P_t \tag{25}$$

These Lotka-Volterra equations (Lotka, 1925; Volterra, 1926) assume that, in the absence of parasites, the host population has Malthusian growth,

$$\frac{dN_t}{dt} = r_n N_t \tag{26}$$

In Equations (25) r_n is the intrinsic rate of increase of the host which is diminished (instantaneously) by a linear proportionality to P_t, the parasite population at time t. Assuming that the parasites die in the absence of hosts, they diminish at a rate $-r_p$ which is offset by a reproductive factor, c_2, per individual of the N hosts attacked.

Consider the ratio

$$\frac{dN_t}{dP_t} = \frac{(r_n - c_1 P_t) N_t}{(-r_p + c_2 N_t) P_t} \tag{27}$$

which reduces to

$$r_p \frac{dN_t}{N_t} - c_2 \, dN_t + r_n \frac{dP_t}{P_t} - c_1 \, dP_t = 0 \tag{28}$$

and integrates to

$$r_p \ln N_t - c_2 N_t + r_n \ln P_t - c_1 P_t = \text{constant} \tag{29}$$

Equation (29) represents a series of closed loops relating P_t to N_t (*Figure 8.3*) in which the constant is determined solely by the starting values of the two populations $N(0)$ and $P(0)$. The equilibrium when $dN_t/dt = dP_t/dt = 0$ is

$$N = \frac{r_p}{c_2} , \qquad P = \frac{r_n}{c_1} \tag{30}$$

Thus, when the populations are disturbed from the equilibrium state they do not return but oscillate about the equilibrium values (*Figure 8.4*). The oscillations then continue indefinitely with constant amplitudes, determined by the starting values of the populations, but with means equal to the equilibria

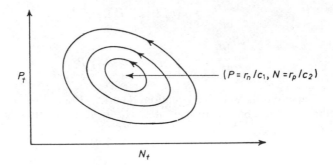

Figure 8.3 Course of cycles of parasite and host population as described by Lotka-Volterra equation with either species self-regulation

$$\frac{1}{T} \int_{t_0}^{t_0 + T} N_t dt = \frac{r_p}{c_2}$$

and (31)

$$\frac{1}{T} \int_{t_0}^{t_0 + T} P_t dt = \frac{r_n}{c_1}$$

We can provide more realism by introducing logistic components into the Lotka-Volterra equations. Simply considering the competition between hosts, if we write

$$\frac{dN_t}{dt} = (r_n - c_1 P_t - bN_t) N_t \tag{32}$$

then the plot of P_t on N_t becomes a closing spiral converging on the equilibrium point (*see Figure 8.5*)

$$N = \frac{r_n c_2}{r_p c_1}, \qquad P = \frac{r_n}{c_1} \tag{33}$$

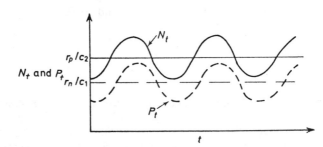

Figure 8.4 Course of parasite and host populations predicted by Lotka-Volterra equations without intra-specific density dependence. Note: parasite population lags behind host. Oscillations show constant amplitude around equilibrium values

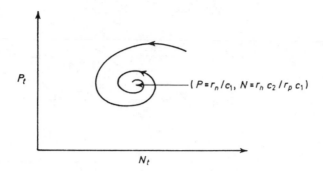

$(P = r_n/c_1, \; N = r_n c_2/r_p c_1)$

Figure 8.5 Course of parasite and host population cycles as described by Lotka-Volterra equations modified to include host self-regulation by density dependent function

to which the populations return after displacement; the trends of the two populations show damped oscillations (*Figure 8.6*).

A simple model can be based on the Lotka-Volterra equations using quite elementary biological assumptions. If we suppose that P parasites each have a complement of F eggs then the mean number of eggs available for oviposition in N hosts will be FP/N. Under the simplifying assumption that each parasite lays an egg completely at random among the N hosts then from the Poisson distribution the probability of a host not being attacked, $P_r(0)$, will be

$$P_r(0) \;\; = \;\; \exp\left(\frac{-FP}{N}\right) \qquad (34)$$

and the two populations at any time $t + 1$ are

$$N_{t+1} \;\; = \;\; f \, N_t \, \exp\left(\frac{-FP_t}{N_t}\right) \qquad (35)$$

and

$$P_{t+1} \;\; = \;\; N_t \left\{ 1 - \exp\left(\frac{-FP_t}{N_t}\right) \right\} \qquad (36)$$

where f is the number of offspring produced per surviving host (Thompson, 1924). The results rest on the assumption that, although some hosts receive more than one parasite egg, only one parasite develops from each host attacked. There are certainly a number of parasite species which can develop more than one offspring per host and Equation (36) needs to be modified to take account of this. Equally there are insect parasites which can 'recognise' parasitised hosts and will normally only oviposit in unparasitised individuals; oviposition in this case is clearly non-random. Nicholson (1933) and Nicholson

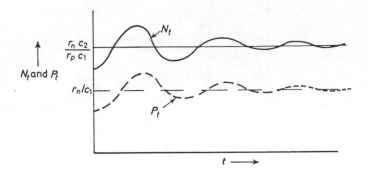

Figure 8.6   Course of parasite and host population predicted by the Lotka-Volterra equations with host self-regulation.   Note: oscillations damp with time and populations converge on equilibrium values

and Bailey (1935) used similar equations in their studies of host-parasite oscillations and of 'steady states'

$$N_{t+1} = fN_t \exp(-aP_t) \tag{37}$$

$$P_{t+1} = N_t \{1 - \exp(-aP_t)\} \tag{38}$$

They interpreted a as the searching power of the parasites (= the area of discovery).   Although they discussed their model in terms of equilibrium values and steady states, it has been shown (Hassell and May, 1973) that models such as theirs have only one equilibrium point at the parasite and host 'steady densities'

$$N^* = \frac{f \ln f}{(f - 1)a}$$

$$P^* = \frac{\ln f}{a} \tag{39}$$

The instability of these models is such that small deviations from N^*, P^* lead to oscillations of increasing amplitude and the eventual extinction of one or other of the two species.   Some measure of density dependence, related to density of the host for instance, will stabilise the system.

When a parasite encounters a host it takes some time to insert the egg into the host and therefore some of its lifetime is used up in the process of oviposition.   This 'handling time', T_h, (Holling, 1959) must be deducted from the total time, T_t, it devotes to search for hosts. The total time spent handling hosts is then linearly related to the number of host encounters it makes by

$$T_s = T_h N \tag{40}$$

Given that the total time, T_t, is finite, the number of hosts attacked cannot increase linearly with host density but increases at a diminishing rate. The 'functional response curve' of Holling defines the number of hosts attacked, N_a, as

$$N_a = P\left(\frac{a' T_t N}{1 + a' T_h N}\right) \tag{41}$$

where a' is an attack coefficient. The host equation is then

$$N_{t+1} = fN_t \exp\left(-\frac{a' T_t P_t}{1 + a' T_h N_t}\right) \tag{42}$$

again assuming a random allocation of eggs among hosts. Hassell and May (1973) show that this model is less stable than the Nicholson-Bailey model; the degree of instability is dependent on f and T_h.

Further insight into the stability of natural populations is given when it is noted that a parasite's searching efficiency can be affected by interference from other individuals of the same species which is proportional to the parasite density (Hassell, 1971). Interference can lead to time being wasted when two adults meet, for example by avoiding each other or by interrupting egg-laying; the interference may be serious enough to force individuals to leave the area in which potential hosts live. An interference component can be included in the Nicholson-Bailey model as a modification of the area of discovery (Hassell and Varley, 1969) such that

$$a = QP_t^{-m} \tag{43}$$

where Q, the quest constant, equals a when $P_t = 1$, and m is the mutual interference constant. The model now becomes

$$N_{t+1} = fN_t \exp(-QP_t^{1-m}) \tag{44}$$

and with handling time

$$N_{t+1} = fN_t \exp\left(-\frac{a' T_t c P_t^{1-m}}{1 + a' T_h N_t}\right) \tag{45}$$

(Hassell and Rogers, 1972). Equation (45) reduces to (42) when $m = 0$ and to (44) when $T_h = 0$. Hassell and May (1973) have discussed the stability properties of these models.

An important basis of the models discussed so far is that they assume random attacks by the parasites with all N hosts equally available, and susceptible, to attack by a parasite. However, it is clear that most animal species are not distributed evenly throughout their full range of habitat; pockets of high density occur in areas which on average may have a relatively low density. The distance a parasite has to travel from one host to another must be of some importance which is certainly not taken into account in the random attack models. It is reasonable to suppose that parasites would modify their behaviour to take advantage of local concentrations of hosts, for example, spending more time or congregating (aggregating) in the higher density area. However, in doing this there is a conflict between the parasite attempting to maximise the number of hosts it can attack and avoiding interference by other individuals of its own species which congregate in the same area. We can imagine a situation where there is a continually changing density as parasites immigrate into the high host density area and then disperse as interference becomes intolerable.

Spatial heterogeneity can be modelled quite simply by assuming n sub-areas each of which has a different host density and then allowing the random attack models to operate in each of the areas (Hassell and May, 1973). The host population at any time $t + 1$ for the simple random model is

$$N_{t+1} = fN_t \sum_{i=1}^{n} \{a_i \exp(-a\beta_i P_t)\} \tag{46}$$

and with interference

$$N_{t+1} = fN_t \sum_{i=1}^{n} [a_i \exp\{-Q(\beta_i P_t)^{1-m}\}] \tag{47}$$

where a_i and β_i are the fractions of the total hosts and parasites respectively in each of the i areas ($i = 1, 2, ..., n$),

$$\sum_{i=1}^{n} a_i = \sum_{i=1}^{n} \beta_i = 1 \tag{48}$$

To simplify calculations define

$$\beta_i = c \, a_i{}^\mu \tag{49}$$

such that

$$c = \left(\sum_{i=1}^{n} a_i{}^\mu\right)^{-1} \tag{50}$$

and μ is the 'parasite' aggregation index. When $\mu = 0$ the parasites distribute themselves evenly over all areas; when $\mu = 1$ then $a_i = \beta_i$ for all i, and when $\mu > 1$ there is differential aggregation of parasites increasing exponentially with host density.

The modifications suggested increase the stability area considerably and depend on a_i, β_i and f (the effective rate of increase of the host). Hassell and May discuss the parameters used in their models in terms of desirable characteristics to seek in parasites potentially suitable for biological control. In effect they have drawn up some of the specifications of the 'perfect' natural control agent. It is up to the biologist to investigate the parameters they have used and then to decide whether they are reasonable or not.

The evolution of the parasite-host models is a specialised example of the approach insect ecologists adopt in quantitative biology. There should be a close marriage between the analytical and experimental methods each suggesting ways in which the other could proceed. The models are still simple descriptions of very complex interactions yet considerable insight into controlling mechanisms has been achieved. Whether increasing the number of parameters will help is a difficult question to answer. The biological information becomes increasingly difficult to obtain and the equations rapidly become unmanageable using standard techniques.

8.4. References

ANDREWARTHA, H.G. and BIRCH, L.C. (1954). *The Distribution and Abundance of Animals.* University of Chicago Press

CONWAY, G.R. and MURDIE, G. (1972). 'Population Models as a Basis for Pest Control', in *Mathematical Models in Ecology,* 195-213. Ed. J.N.R. Jeffers. Oxford University Press; London

EHRLICH, P.R. and EHRLICH, A.H. (1972). *Population, Resources, Environment,* Issues in Human Ecology. Freeman; San Francisco

EMLEN, J.M. (1973). *Ecology: An Evolutionary Approach.* Addison Wesley; Reading, Massachusetts

HASSELL, M.P. (1971). 'Mutual Interference Between Searching Insect Parasites', *J. Anim. Ecol.,* **40**, 473-488

HASSELL, M.P. and MAY, R.M. (1973). 'Stability in Insect Host Parasite Models', *J. Anim. Ecol.,* **42**, 693-726

HASSELL, M.P. and ROGERS, D.J. (1972). 'Insect Parasite Responses in the Development of Population Models', *J. Anim. Ecol.,* **41**, 661-676

HASSELL, M.P. and VARLEY, G.C. (1969). 'New Inductive Population Model for Insect Parasites and Its Bearing on Biological Control', *Nature,* **223**, 1133-1137

HOLLING, C.S. (1959). 'The Components of Predation as Revealed by a Study of Small-mammal Predation of the European Sawfly', *Canad. Ent.,* **91**, 293-320

HOLLING, C.S. (1966). 'The Strategy of Building Models of Complex Ecological Systems', in *Systems Analysis in Ecology,* 195-214. Ed. K.E.F. Watt. Academic Press; New York

HOLLING, C.S. (1968). 'The Tactics of a Predator', in *Insect Abundance,* 47-58. Ed. T.R.E. Southwood. Oxford University Press; London

KREBS, C.J. (1972). *Ecology: The Experimental Analysis of Distribution and Abundance.* Harper and Row; New York

LOTKA, A.J. (1925). *Elements of Physical Biology,* William and Wilkins; Baltimore. [*Elements of Mathematical Biology* (1957). Dover; New York

MALTHUS, T.R. (1798). *An Essay on the Principles of Population.*
London [various editions, e.g. Everyman's Library, 2 vols. (1914);
Economic Classics, New York (1909)]

MAY, R.M. (1972). *Stability and Complexity in Model Ecosystems.*
Princeton University Press; Princeton

MEADOWS, W.H., MEADOWS, D.L., RANDERS, J. and BEHRENS, W.W. (1972).
The Limits to Growth. Universe Books; New York

NICHOLSON, A.J. (1933). 'The Balance of Animal Populations',
J. Anim. Ecol., **2**, 132

NICHOLSON, A.J. and BAILEY, V.A. (1935). 'The Balance of Animal
Populations: Part I', *Proc. Zool. Soc. London,* **3**, 551

PEARL, R. (1927). 'The growth of populations', *Quart. Rev. Biol.,* **2**,
532

PIELOU, E.C. (1969). *Introduction to Mathematical Ecology.* Wiley;
New York

ROYAMA, T. (1971). 'A Comparative Study of Models for Predation
and Parasitism', *Res. popul. Ecol.,* suppl. 1

THOMPSON, W.R. (1924). 'La théorie mathématique de l'action des
parasites entomophages et le facteur du hassard', *Ann. Fac. Sci.
Marseille,* **2**, 69

VERHULST, P.E. (1838). 'Notice sur la loi que la population suit
dans son accroissement', *Corresp. Math. Phys.,* **10**, 113

VOLTERRA, V. (1926). 'Variationi e fluattuazioni del numero
d'individui in specie animali conviventi', *Mem. R. Accad. Naz. dei
Lincei.,* Set. VI, **2a** [English translation in Chapman, R.N., *Animal
Ecology,* Ed. R.N. Chapman. McGraw-Hill, New York]

de WIT, C.T. (1960). 'Space Relationships within Populations of One
or More Species', *Symp. Soc. Exptl. Biol.,* **15**, *Mechanisms in
Biological Competition,* 314

8.5.   Problems for further study

1.   Consider the logistic equation

$$N_t = \frac{N_{eq}}{\{1 + a \exp(-kt)\}}$$

Show that the maximum rate of growth occurs at the point $1/k \ln a$
and that the population size at that time is $N_{eq}/2$.

The Gompertz growth curve is similar to the logistic, i.e.

$$N_t = N_{eq} \, e^{-be^{-kt}}$$

What is the point of inflexion on this curve and what is the value of
N at this point?

2. Using the Hassell-Varley model (44) investigate the parasite and host population fluctuations putting $Q = 0.20$; $f = 2.0$; $P(O) = 10$ and $N(O) = 50$ for

(a) $m = 0$ ($\equiv$ no interference $\equiv$ Nicholson-Bailey model, Equations (37,38))
(b) $m = 0.2$ (weak interference)
(c) $m = 0.6$ (strong interference)

Laboratory experiments with the pea aphid (greenfly) gave the following data:

Age (days)	Large individuals		Small individuals	
	l_x	m_x	l_x	m_x
0-4	1.0	0	1.0	0
5-8	1.0	0	1.0	0
9-12	1.0	24.04	1.0	15.00
13-16	0.978	29.94	0.962	20.88
17-20	0.913	23.34	0.889	19.42
21-24	0.783	10.00	0.796	14.10
25-28	0.522	2.94	0.648	6.28
29-32	0.370	0.36	0.574	1.74
33-36	0.217	0	0.370	0
37-40	0.130	0	0.111	0

Size is said to influence the rates of increase of this aphis species. Determine the intrinsic rates of increase (λ_1) for the two groups and the stable age distribution. Use both the life-table and Leslie matrix methods (Leslie (1945), *Biometrika,* **35**, 183-212; Leslie (1948), *Biometrika,* **38**, 213-245).
Note the post-reproductive period which is characteristic of this species.

3. Parasite-host models based on the Lotka-Volterra Equations (25) have typical oscillatory behaviour. What is the time period of these oscillations and is it the same for both parasites and hosts? By how much does the parasite population, e.g. measured at the peaks, lag behind that of the host?

9

A DIFFERENTIAL MODEL OF DIABETES MELLITUS

M. J. Davies
Applied Mathematics Department, University College of Wales, Aberystwyth

[Prerequisite: elementary differential equations]

9.1. Introduction

The disease known as diabetes mellitus is well known. It is a condition in which certain processes of the body concerned with the metabolism of sugar to energy fail. Blood sugar concentrations rise without the appropriate mechanisms which regulate the amount of sugar in the blood stream performing properly. The principle homeostatic influence is that of insulin which is involved in the sugar metabolism. One way or another it is this mechanism which fails.

In general, differential models are associated with those naturally occurring processes in which rates of change of variables are involved. That is to say, one has a dynamic situation with various quantities changing continuously with time. Observation or experiment reveals the operational statements that one can make about the changes. These are displayed as equations involving the variables themselves and their derivatives, namely differential equations. More involved processes can lead to more complex models, but for the present, attention is directed at obtaining a simple differential model of diabetes.

9.2. Variable identification

The two main variables involved will be the obvious quantities which one can observe or manipulate clinically; these are the blood sugar level, x, and the blood insulin level, y. Two additional minor variables also play their part, the food input, z, and for the diabetic person, the insulin input, w.

9.3. State relations

Briefly, the qualitative biochemistry of a normal person is as follows (Campbell, Dickenson and Slater, 1963). The stable reference state is a blood sugar fasting level, x_0, with zero blood insulin level. Levels not at the stable point cause changes in both one and the other by a number of separate mechanisms.

1. If the blood sugar rises above its fasting level, insulin is secreted by the pancreas into the bloodstream. One can model this in a piecewise linear manner by writing

$$\left[\frac{dy}{dt}\right]_{(i)} = b_1(x - x_0), \qquad x > x_0$$

$$= 0, \qquad\qquad x < x_0$$

2. Insulin itself degrades by separate biochemical processes, half the free insulin *in vivo* being rendered inactive in about 10-25 minutes. Hence one can write

$$\left[\frac{dy}{dt}\right]_{(ii)} = -b_2 y, \qquad y \geqslant 0$$

3. Any external source of insulin will be an explicit forcing term in the differential equation. In the non-diabetic, this source term will be identically zero, for the diabetic it will be a function of time representing the pattern of injections.

$$\left[\frac{dy}{dt}\right]_{(iii)} = b_3\ w(t)$$

The three constants b_1, b_2 and b_3 are all positive, by construction. They can be called sensitivities and are respectively the sensitivity of the insulin gradient to (a) high blood sugar levels (b) insulin levels, and (c) the input. If one uses the step function defined by

$$H(\xi) = 0, \qquad \xi < 0$$

$$= 1, \qquad \xi \geqslant 0$$

to absorb the complication in (a), the total gradient of the insulin level can now be written as

$$\frac{dy}{dt} = b_1(x - x_0)\, H(x - x_0) - b_2 y + b_3 w(t)$$

In the case of the sugar level gradient, the following components can be identified:

1. The presence of insulin induces metabolism of the sugar, thus reducing the blood sugar level. The higher the blood sugar level or the insulin level, the faster is the reduction. This suggests that for small variations at least the product of the two levels will be an adequate representation.

$$\left[\frac{dx}{dt}\right]_{(i)} = -a_1 xy$$

2. If the blood sugar level drops below its fasting level, (say because of physical activity when hungry) sugar is released from the liver 'stores' to raise the level back to normal.

$$\left[\frac{dx}{dt}\right]_{(ii)} = a_2(x_0 - x), \qquad x < x_0$$

$$= 0, \qquad x \geqslant x_0$$

3. There is a very small 'natural' decay of sugar concentrations, whose effect in operation of the model is minor. Hence, although it is in fact built in, no further consideration of it will be made and it will be ignored in subsequent discussion.

$$\left[\frac{dx}{dt}\right]_{(iii)} = -a_2'(x - x_0)$$

4. The external source of blood sugar is the normal feeding pattern. This is represented by an explicit function of time

$$\left[\frac{dx}{dt}\right]_{(iv)} = a_3 z(t)$$

Again the constants a_1, a_2, a_2', and a_3 are positive and are respectively the sensitivity of the sugar level gradient to (a) the presence of insulin (b) low blood sugar levels (c) high blood sugar levels and (d) the input. Using the same step function, the total gradient of the blood sugar level can be written as

$$\frac{dx}{dt} = -a_1 xy + a_2(x_0 - x)H(x_0 - x) - a_2'(x - x_0)H(x - x_0) + a_3 z(t)$$

Notice that the variation of blood sugar level above or below the fasting level brings two different stabilising processes into play, so that the system is essentially non-linear. In addition the principal process which removes sugar depends on the presence of both sugar and insulin; this sink term is therefore again basically non-linear, and the representation used, a product, is probably the most simple conceptually.

It should be emphasised that this model is technically the most elementary one that could possibly represent the realisation. The various terms are piecewise linear with the exception of the interaction, which is a product of linear terms. In fact in the real situation, peaking, cut-off and hysteresis features are present and the suggested model is correspondingly rather crude.

9.4. The source terms

The normal input to the blood sugar level is via a food source rather than a direct intervention with the blood sugar level itself. This food store is filled periodically rather than continuously, and it will be taken that the contents at any stage are depleted in a simple exponential manner. The actual sugar source term can then be written as $a_3 z(t)$ where

$$z(t) = 0, \qquad\qquad t < t_0$$
$$= Q e^{-K(t - t_0)}, \qquad t \geqslant t_0$$

The information to be given about each meal will thus be Q (quantity), K (delay parameter), t_0 (time of the meal). A variation of K can accommodate various types of food input. Simultaneous large values of Q and K together could simulate an injection of glucose solution. The parameters a_3 and Q of course appear only as a product, but it is convenient to separate them merely to get the 'quantity' of a meal measured by a numerical value which sounds familiar.

The normal input of insulin is the closed loop feedback previously described. However, in the case where this mechanism is faulty, an input similar to the sugar input is employed. A sub-cutaneous injection at periodic intervals can be modelled as filling up an 'insulin store' which leaks its contents into the system over a period of time. One knows that the 'maximum effect' of an injection is felt after a certain lag (normally about 3 h) and that the effect has petered out after some total time. If one takes 'maximum effect' to mean maximum leaking rate and consequently maximum effect on the insulin gradient, one can model this description by taking $w(t)$ to be the function shown in *Figure 9.1*. The formula as a piecewise function of time can be quite easily written down, and requires the following data: time of injection, quantity of injection, lag time to maximum, and the slopes up and down of the ramps. Again the parameter b_3 is introduced – insulin sensitivity to injection – in order that the quantity of an injection is measured by a numerical value of familiar size.

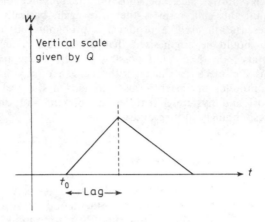

Figure 9.1   Insulin source function w(t)

9.5.   The analysis

The presence of the product term xy means that no elementary analysis is possible.   The piecewise linear linkages and the various source terms present no essential difficulty, and were it not for the product term, some algebraic formulae could be developed.   However, their use would be limited since all the modelling has grossly simplified the interactions even for the two-variable case.   All one really wants to know are the overall features of the response of the system to various inputs, and for this, pictures suffice.   A simple program was written to solve the equations numerically and to display the results graphically on the line printer.

9.6.   Discussion

Figure 9.2 shows the sugar-insulin response of the model for a normal person (i.e. one not suffering from diabetes) to a day of three meals. At any meal time the blood sugar level rises and stimulates the pro- duction of insulin; the presence of insulin in its turn stimulates the depression of the blood sugar level and is itself removed by its

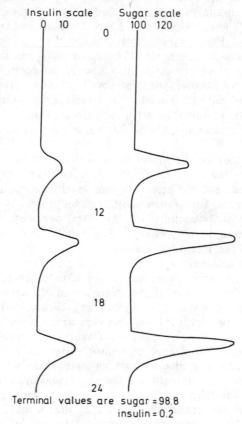

Figure 9.2 *Normal sugar-insulin response*

natural decay process. The numerical data for this model are as follows:

Sensitivities: a_1 = 0.05
                a_2 = 1.0
                a_3 = 4.0
                b_1 = 0.5
                b_2 = 2.0
                b_3 is immaterial

Meals: Breakfast at 08.00 hours of 50 units
                Lunch at 13.00 hours of 100 units
                Dinner at 20.00 hours of 100 units.

Some comments about these numbers would be here appropriate. It has been previously pointed out that the value of a_3 is tied to the quantity of the meal, hence these are to some extent arbitrary. One could pick a_3 = 1 with meal sizes four times larger without any real change in the model. The constant b_2 determines the rate of decay of free insulin *in vivo*; the choice $b_2 = 2$ gives insulin a half-life of about 20 min. The choice $b_2 = 3$ would yield a half-life of about 12 min. Hence experiments indicate that the actual value of b_2 is somewhere in this range. The constant a_2 measures the rate of recovery of the model from a low blood sugar, its value of unity gives a half-recovery time of about 40 min. The constants a_1 and b_1 are more intimately related, and their given values were chosen by trial and error in order to get a realistic simulation. The constant b_3 is of course immaterial for this case since no external input of insulin was used.

This part of the model building problem, hitherto unmentioned, is probably the most difficult in practice. The state variables have been identified, and the state relations established with a kernel system abstracted. The *system identification* process is now the technical exercise of determining the numerical value of the various parameters that appear in the system and is usually the most time consuming part of the modelling exercise.

The parameters a_1, a_2, ..., b_1, b_2, ... , etc. are determined by comparing the results from the model with observation; such parameters can be found only if the structure of the model is realistic. The acid test of a model, or indeed any theory, is prediction; if the system can be identified only retrospectively then it is surely of doubtful virtue. Hence, this model is of value if it enables us to make predictions beyond the range of known observations.

One can now identify two separate 'diabetic' effects in the model: these are reductions in a_1, the sugar sensitivity to insulin, and b_1, the insulin sensitivity to sugar. The first of these can be interpreted in terms of the commonly found fact that some diabetics need far higher insulin levels (reflected in terms of the size of their injections) than does the normal person. The second corresponds to the more directly visualised depression, or indeed complete absence, of the insulin production term.

In either or both of these cases consistently high blood sugar levels are produced and a host of other factors and processes, ignored in the model, come into play. As far as the model is concerned, reduction of the sensitivities a_1 and b_1 will render the system unstable and it will no longer seek strongly its stable fasting state. To compensate for this, the insulin source term $w(t)$ is introduced. It is illuminating to note that what one has here is a sugar-insulin system stable to any $z(t)$ input in the normal person. The diabetic problem is that the system has become unstable. Mild instabilities, that is, mild diabetes or minor alterations in the salient sensitivities, can be controlled by appropriate choice of $z(t)$ alone. This is classic dietary control and cannot be represented on the present model because of its bare simplicity, as mentioned earlier. Severe diabetes necessitates the introduction of a second controller $w(t)$ before stability can be restored.

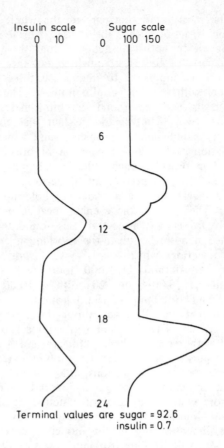

Insulin scale
0   10

Sugar scale
0   100 150

6

12

18

24

Terminal values are sugar = 92.6
insulin = 0.7

Figure 9.3   Controlled diabetic sugar-insulin response

The diabetic model, whose sugar-insulin response is shown in *Figure 9.3,* was characterised by the following changes in the system. The sugar sensitivity to insulin a, was reduced from 0.05 to 0.03, while the insulin sensitivity to sugar was almost completely eliminated, b_1 being changed from 0.5 to 0.01. The input sensitivity b_3 proved to be realistic with a value of 1.0.

The whole point is now to choose judiciously an insulin input $w(t)$ in such a way that the resulting open-loop system, and the sugar-insulin response to the habitual food intake, is mimicked to some degree of closeness. Some latitude is permissible and what one seeks is a daily insulin input which, while changing to some extent the normal sugar-variation during the day, nevertheless keeps the perturbation within limits and also renders the whole system periodic over a space of many days.

By far the most interesting part of the whole problem now appears. The criterion is adopted that the classic clinical pattern of two injections per day is to be established and that these need to 'cover' the three main meals of breakfast, lunch and dinner. It must first be realised that it is impossible to mimic the three peak pattern of the normal model with but two insulin inputs. The first injection must cover two meals, and the second the third meal – and any residue from the first two. The remarkable fact that emerges is that even with this very simple model, problems raise which closely parallel clinical situations. The time separation of breakfast and lunch must be shortened in order to enable the morning injection to cover both meals. In addition, to prevent an 'undershoot' between the meals, a mid-morning snack must be provided to hold up the plunging blood sugar level. The time of the evening meal is pulled back to 18.00 hours merely for convenience, and it is found that the evening injection needs to be much smaller than the morning injection.

The two injections which gave rise to *Figure 9.3* were 20 units at 08.30 hours and 14 units at 18.00 hours, while the meals they covered were 50 units at 08.00 hours, 20 units at 10.30 hours, 50 units at 12.00 hours and 100 units at 18.00 hours.

Many comments can be made about this model, and its reflection of the realisation. The first and really remarkable point is that a two state model with piecewise linear linkages and a simple product interaction can represent crudely yet faithfully the real medical situation: timing of the meals and injections, the mid-morning snack etc. The model could be improved in two major ways. The interactions could be made more complex by the introduction of thresholds, peaking, and other non-linear features. Probably of more value would be to increase the dimensionality of the model, one could even comment that the model will be unrealistic until this is done. Consider the source and sink terms for the sugar level. Sugar is metabolised for various purposes: energy production, deposition in tissues, and repletion of the liver store. The source of sugar is the input, a 'fast' liver store and a 'slow' tissue store. Now the liver store is finite, in fact it holds only about 250 g of glucose. Hence, if one wants a model to reflect a continuing situation, one of its variables must be the liver store content, its response to low blood sugar level being non-linear to that level and most certainly non-linear to its own content. In a similar vein the slow tissue store of blood sugar can be modelled, identified as the weight of the person, the various sensitivities then possibly depending in their turn upon that weight. The complexities are legion, and the more one thinks about the problem, the more interesting it becomes.

One of the major arguments in favour of the given simple model is that its input-output information, visualised with a black box in the middle, is very near indeed to clinical practice. The principal output is blood sugar level, and the whole system is stabilised by dietary and insulin control. The glucose tolerance test yields three values of the blood sugar level at ½, 1 and 2 hours following a standard meal input of some 100 g of glucose taken at the stable fasting state, and serves to differentiate between normal and diabetic responses. Finally, the response of the black box to early treatment conditions further treatment regimes. It almost goes without saying that the condition of the black box itself plays a major part in the exercise.

The crux of this problem is fairly obvious. The sugar metabolism system – whatever it may be – is a strongly homeostatic process. Two key parameters can be identified, the sugar and insulin sensitivities each to the other. Hence any model which includes these will represent the true situation with some degree of accuracy, and this is exactly what the present two-state model does. It is the technically most simple model of lowest order which includes the key sensitivities.

9.7. References

CAMPBELL, E.M.J., DICKENSON, D.J. and SLATER, J.D.R. (1963). 'Clinical Physiology', Chapter 12 in *Energy Sources and Utilization*, Blackwell Scientific Publications; Oxford

DAVIS, H.T. (1962). *Introduction to Non-Linear Differential Equations*. Dover; New York

9.8. Problems for further study

1. Find the equation connecting b_2 with the insulin half-life.

2. Model the insulin secretion term to include saturation (i.e. a maximum rate of secretion) (a) in a piecewise linear manner; (b) smoothly.

3. The source term for sugar, $z(t)$, is actually the response of a sub-system to a delta function input. What is this system?

4. The source term for insulin, $w(t)$, is actually a rough piecewise representation of a displaced Gaussian input; what would be the formula for this? What actual sub-system would respond to what input in a like manner?

5. Append to the system a 'fast' liver store with contents $\lambda(t)$. This will be the source of the sugar liberated by low blood sugar levels. Write down the equation for $\dot{\lambda}(t)$ introducing the linkages which deplete *and restock* the liver store. (Note: These will be non-linear since if $\lambda = 0$, then no sugar can be released.)

6.   Apart from the cut-off features represented by the Heaviside function, the structure of this problem is also that of one known as the foxes and rabbits problem considered by Volterra (*see* Davis, 1962). Find out what that problem is and compare the roles played by the variables.

10
STOCHASTIC MODELS FOR ROAD TRAFFIC SITUATIONS

Winifred D. Ashton
Department of Mathematics, University of Surrey

[Prerequisites: elementary probability theory and the Laplace transformation]

10.1. Models for road traffic

The problems presented by the growth of modern road traffic are both acute and urgent. From the viewpoint of the community they have considerable social impact, whilst to the mathematician they present a challenging area for the application of theoretical methods covering a wide range of techniques.

Areas which are amenable to mathematical investigation include the following:

1. Movement of vehicles at intersections and on the open road, and parking problems.
2. Engineering problems, such as the design of road systems for new towns and the investigation of systems of vehicular control including traffic lights and roundabouts.
3. Transportation and scheduling problems for fleets of vehicles, together with a study of their economic consequences.
4. Accident and safety aspects.

Methods of attack on these problems have covered a wide range. One approach for dense flow involves treating the flow of vehicles down a road in kinematic terms as a fluid (Lighthill and Whitham, 1955). In this case solutions are effected using classical methods. These models are macroscopic in that the properties of individual vehicles do not enter, so such theories cannot be expected to explain traffic behaviour in any detail. Dynamical or car-following models, on the other hand, are microscopic in the sense that attention is focussed on the behaviour of two consecutive vehicles moving down a road. Both of these types of model are *deterministic* in nature.

A quite different kind of model from the two already mentioned is the *probabilistic* or *stochastic* model in which a random element is incorporated. In the situations considered in the following sections, traffic flow is considered as a stochastic point process, the probabilistic structure of which is to be inferred as far as possible from data collected in real situations. Such formulations, involving statistical variables, need sophisticated mathematical techniques to exploit them.

All these types of model have their own area of application and they can be used to implement one another. Some theoretical approaches have resulted in formulations which have been criticised for their inadequacy or artificiality. Vehicles are often, for example, regarded as geometrical points. In the traffic field it seems peculiarly difficult to find a theoretical approach which is both realistic and sufficiently tractable to be fruitful. As a practical alternative, therefore, recourse must often be had to simulation, or perhaps a combination of theory and simulation.

10.2.   A simple model for pedestrian delay

Consider the following simple situation. A pedestrian wishes to cross a road along which traffic is flowing at a given rate. Assume that the traffic is flowing in a single lane, that the pedestrian requires a given minimum time to complete the crossing and that he arrives at the crossing point just as a vehicle passes. Some of these restrictions can be removed, as will be seen later. The situation can be depicted diagrammatically as in *Figure 10.1.*

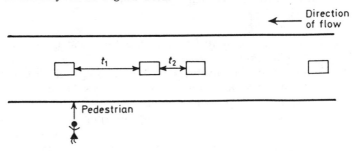

Figure 10.1   Model for pedestrian delay

Quantities of interest in this problem include the probability of the pedestrian being delayed and the length of any such delay incurred. Firstly, however, the model must be made more specific — we need a set of rules governing the behaviour of both pedestrian and vehicles.

Consider the pedestrian. Clearly he bases his decision to cross the road on a number of factors, the most important one of which involves the proximity and speed of the nearest vehicle. These factors can, say, be accounted for by considering the time-gap currently faced by the pedestrian, i.e. the time that the nearest vehicle will take to reach him. In delay problems, time gaps rather than distance gaps are important. Time gaps between successive vehicles are termed *headways,* and are usually measured from front bumper to front bumper.

The pedestrian's decision criterion for crossing must now be specified. It might be assumed, for instance, that each individual has a so-called *critical gap,* that varies from individual to individual and from situation to situation, but that any one individual in a given situation always rejects gaps less than his critical gap and always accepts longer ones. This is the simplest model for gap-acceptance that can be set up, and is represented by the step-function, $\Gamma(t)$, illustrated in *Figure 10.2.*

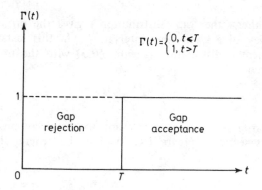

$$\Gamma(t) = \begin{cases} 0, & t \leqslant T \\ 1, & t > T \end{cases}$$

Figure 10.2   Step-function for gap-acceptance

This model accords fairly well with observation, but other models are possible.   Some of these are suggested in Section 10.6.

Now consider the way vehicles are distributed (in time) along the road.   In stochastic models it is not specified how any particular vehicles are placed on the road.   There is, instead, a probability distribution for the headways between vehicles in a single line of traffic.

The simplest possibility is that headways are 'random'.   In statistical theory this word does not mean 'vaguely haphazard' but something quite specific.   A series of events, in this case the arrival times of a series of vehicles at a point on the road, is said to be random when:

1. The probability of an event in a small time interval $(t,\ t + \delta t)$ is $\lambda \delta t + O(\delta t)$, where λ is a constant.
2. The probability of two or more events in $(t,\ t + \delta t)$ is $o(\delta t)$.
3. The number of events in $(t,\ t + \delta t)$ is independent of what has happened in $(0,t)$.

The constant λ defines the rate, i.e. the mean number of vehicles per unit time and $o(\delta t)$ indicates a small-order quantity which can be neglected.   Note that for dense traffic both (2) and (3) break down, and the hypothesis becomes unrealistic (*see* Problem (5)). It is not surprising, therefore, that the distributions which are immediately derivable from the random hypothesis stated above do not describe empirical data accurately for flows in excess of 800 v.p.h. in one lane, i.e. when the mean time gap is less than 4.5 s.

Two distributions, which equivalently describe random flow, can be derived from the hypothesis. The 'counting distribution' gives the probability of any given number of vehicles arriving at a point in a unit time interval. It turns out here to be the Poisson distribution: if $P(k)$ is the probability of k arrivals in unit time, then

$$P(k) = \frac{e^{-\lambda} \lambda^k}{k!}, \qquad k = 0, 1, 2, \dots$$

The other — the 'gap' distribution — gives the probability of a gap or headway of a given time interval, t. In this instance this is the Exponential distribution (*Figure 10.3*) with the probability density function

$$f(t) = \lambda e^{-\lambda t}, \qquad t > 0$$

The latter distribution is relevant to the delay problem posed earlier in this section. *Figure 10.3* depicts the general shape of the distribution.

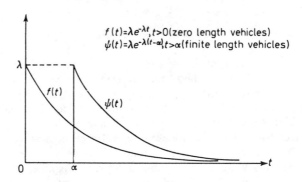

Figure 10.3   The Exponential and displaced Exponential distributions

The probability of encountering a gap greater than a given time interval, T, is given by

$$Pr \ (\text{gap} > T) = \int_T^\infty \lambda e^{-\lambda t} \ dt = e^{-\lambda T}$$

For a flow-rate of 600 v.p.h., say, $\lambda = 1/6$ v.p.s. If the pedestrian requires, say, a gap of at least 8 s in which to cross the road, then the probability of his being able to cross without delay is $e^{-8/6} = 0.2636$. If the pedestrian has to wait for the first vehicle to pass, he repeats the decision process with the second gap, and so on. The actual delay incurred is then obtained by summing random gaps. This is dealt with in the next section.

Before proceeding let us consider how the model can be made more realistic. To begin with, since vehicles have a finite length, the headway between two vehicles is always non-zero. Thus a displaced Exponential distribution is more appropriate (*see Figure 10.3*). If the minimum time gap is a, where a is probably of the order of one second, the required distribution is

$$\psi(t) = \lambda e^{-\lambda(t-a)}, \qquad t > a$$

The mean headway of the Exponential distribution (for which $a = 0$) is easily shown to be $1/\lambda$, for the displaced Exponential distribution it is $(1/\lambda + a)$.

The restriction in the model which ensures that a pedestrian's arrival coincides exactly with the passage of a vehicle can also be removed. A well-known result, applicable for the (undisplaced) Exponential distribution only, states that the residual gap from the pedestrian's arrival to the arrival of the next vehicle, termed a *lag*, has exactly the same distribution (with the same mean) as a complete gap. In delay problems, distributions are required both for the lag, dealing with the first interval encountered, and for gaps, dealing with all subsequent ones. Only in the case of random flow are lags and gaps on the same footing. In other cases the analysis is more complicated.

The restriction that traffic moves in a single lane may also be lifted. If there are two random streams with flow-rates, λ_1 and λ_2, moving *independently* in the same direction, then it can be shown very simply that gaps in the combined streams are given by the probability density function

$$\phi(t) = (\lambda_1 + \lambda_2) e^{-(\lambda_1 + \lambda_2)t}, \qquad t > 0$$

For this stream, the mean gap is $1/(\lambda_1 + \lambda_2)$, and the probability of a pedestrian arriving in a gap of length greater than T is

$$e^{-(\lambda_1 + \lambda_2)T}$$

This result can be extended to more than two random streams, and to other distributions.

One point should be noted here. If the pedestrian is to cross a road having more than one stream of traffic, he may do so in at least two ways. He may consider gaps in the combined stream, decide and cross in one go, or he may consider the streams separately — in which case there is a notional (or perhaps actual) island in the middle of the road. The results are different for the two models.

The case of two streams of traffic, one in each direction, is intractable, unless an island is postulated, in which case the model and results are essentially the same as those for two streams in the same direction.

10.3.   A simple model for a priority intersection

Intersections are important features of the road system, delays at
which can cause impatience and accidents.   They can be controlled in
at least three ways, e.g. by traffic lights, roundabouts or GIVEWAY
signs.   The first two are usually used when the junction is comprised
of roads of equal importance, the third when the junction is between
a major and a minor road.   An intersection incorporating a GIVEWAY
sign on the minor road is usually termed a *priority intersection.*

In this section we shall consider a very simple model for such a
junction (*Figure 10.4*) in which a single driver is waiting at the
GIVEWAY sign on the minor road for an opportunity to cross the
major road stream, which is taken to be random and in one lane.
This problem is essentially the same as that of a pedestrian crossing a
road, but in this section the model will be described in terms of the
priority intersection.

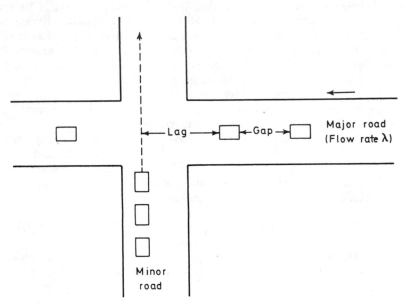

Figure 10.4   Priority intersection

It should be noted that answers from this kind of model must be
stochastic.   In this case it takes the form of a distribution for the
delay to the vehicle in the minor road, and yields, in particular, an
expression for the mean delay.

In order to obtain an explicit solution we shall need the Laplace
transformation.   Let $f(t)$, $F(t)$ and $f^*(s)$ be, respectively, the probability
density function, distribution function and Laplace transform of the
major road flow, so that

$$F(t) = \int_0^t f(t) \; dt$$

and

$$f^*(s) = \int_0^\infty e^{-st} f(t) \; dt$$

Note that for random flow

$$f(t) = \lambda e^{-\lambda t}, \qquad t > 0$$

$$F(t) = 1 - e^{-\lambda t}$$

$$f^*(s) = \lambda/(\lambda + s)$$

Before proceeding, a standard result relating to sums of random variables is quoted. Consider two gap distributions, characterised by $f(t)$, $F(t)$ and $f^*(s)$ in the first case and by $g(t)$, $G(t)$ and $g^*(s)$ in the second. The distribution function for the sum, u, of two gaps, one from each distribution, is

$$H(u) = \int_0^\infty \int_0^{u-t} f(t') \; g(t) \; dt' \; dt$$

$$= \int_0^\infty F(u - t) \; g(t) \; dt$$

Differentiation with respect to u under the integral sign gives the probability density function for the sum as

$$h(u) = \int_0^u f(u - t) \; g(t) \; dt$$

We say that h is the *convolution* of f and g and write $h = f*g$. It can be shown that the respective Laplace Transforms are related by

$$h^*(s) = f^*(s) \cdot g^*(s)$$

If f and g are the same distribution, i.e. if gaps from the same distribution are being added, these relations can be written

$$h = \{f\}^{2*}$$

and

$$h^*(s) = \{f^*(s)\}^2$$

Suppose that the minor road vehicle arrives at the GIVEWAY sign at time $t = 0$ and requires a time T to cross the major road. Let the delay distribution relating to the minor road vehicle have density function $w(t)$ with Laplace transform $w^*(s)$.

If the first lag or gap $t_1 > T$, then there is no delay, and $w(t)$ thus has a discrete component

$$Pr\ (t_1 > T)\ =\ 1 - F(T)\ =\ e^{-\lambda T}$$

For those cases where $t_1 < T$, $t_2 < T$ etc., we must consider a continuous component in the range $(0,\infty)$. For this the argument is as follows. If the vehicle must wait until the time gap t_{n+1}, say, where $n > 0$, then the required delay is the sum of the n random gaps,

$$\sum_{j=1}^{n} t_j$$

As the vehicle has already rejected gaps $t_1, ..., t_n$ (all of which are less than T), the density of the t_j th gap $(j \leqslant n)$ is not $f(t)$ as defined earlier, but $f(t)$ normalised over the interval $(0,T)$, i.e.

$$\frac{f(t)}{F(T)}\ =\ \frac{\lambda e^{-\lambda t}}{1 - e^{-\lambda T}}\ ,\qquad 0 < t < T$$

Thus the continuous part of $w(t)$, conditional on the crossing being in the $(n + 1)$th gap, is given by the n-fold convolution of $f(t)/F(T)$, which can be written as $\{f(t)/F(T)\}^{n*}$. To obtain the *unconditional* distribution of $w(t)$ this expression must be multiplied by the probability of crossing in the $(n + 1)$th gap, viz. $\{F(T)\}^n \{1 - F(T)\}$ and the result summed over all n. Adding the discrete component obtained earlier, we obtain the unconditional distribution as

$$w(t)\ =\ \{1 - F(T)\}\ \delta(t) + \{1 - F(T)\} \sum_{n=1}^{\infty} \{f(t)\}^{n*}$$

where $\{f\}^{0*} = 1$ and $\delta(t)$ is the Dirac delta function, i.e.

$$\delta(t)\ =\ \begin{array}{l} \infty,\ t = 0 \\ 0,\ \text{elsewhere} \end{array}$$

and

$$\int_{-\epsilon}^{+\epsilon} \delta(t)\ \mathrm{d}t\ =\ 1$$

Taking the Laplace transform of both sides, we have

$$w^*(s)\ =\ \{1 - F(T)\}\ +\ \frac{\{1 - F(T)\}\int_0^T e^{-st}\ f(t)\ \mathrm{d}t}{1 - \int_0^T e^{-st}\ f(t)\ \mathrm{d}t}$$

$$=\ \frac{1 - F(T)}{1 - \int_0^T e^{-st}\ f(t)\ \mathrm{d}t}$$

For the Exponential distribution $f(t) = \lambda e^{-\lambda t}$, this becomes

$$w^*(s) = \frac{(\lambda + s)\, e^{-\lambda T}}{s + \lambda e^{-(\lambda + s)T}}$$

This function is difficult to invert but can be shown to be

$$w(t) = e^{-\lambda T}\, \delta(t) +$$

$$\lambda e^{-\lambda T} \sum_{j=0}^{r-1} \left[(-e^{-\lambda T})^j \left\{ \frac{[\lambda(t - jT)]^{j-1}}{(j-1)!} + \frac{[\lambda(t - jT)]^j}{j!} \right\} \right]$$

$$(r - 1)\, T \leqslant t \leqslant rT, \qquad r = 1, 2, \ldots$$

Fortunately, if only the mean and variance of the delay distribution $w(t)$ are needed, the inversion need not be carried through. A standard result (e.g. Hogg and Craig, 1970, p.49) gives the formulae below:

$$\text{mean} = -\left.\frac{\mathrm{d}}{\mathrm{d}s}\, w^*(s)\right|_{s=0}$$

$$\text{variance} = \left.\frac{\mathrm{d}^2}{\mathrm{d}s^2}\, w^*(s) - \left(\frac{\mathrm{d}}{\mathrm{d}s}\, w^*(s)\right)^2\right|_{s=0}$$

For this case these yield

$$\text{mean} = \frac{1}{\lambda}\left(e^{\lambda T} - \lambda T - 1\right) \qquad (1)$$

$$\text{variance} = \frac{1}{\lambda^2}\left(e^{2\lambda T} - 2\lambda T e^{\lambda T} - 1\right)$$

If the discrete component is omitted, the Laplace transform for *delayed* vehicles is

$$\frac{\lambda e^{-\lambda T}\{1 - e^{-(\lambda + s)T}\}}{s + \lambda e^{-(\lambda + s)T}}$$

Differentiation then gives the mean delay as

$$\frac{e^{\lambda T}}{\lambda} - \frac{T}{1 - e^{-\lambda T}} \qquad (2)$$

If it can be assumed that $\lambda T < 1$, the exponential in the Expression (1) can be expanded to give, approximately,

$$\text{mean delay} \simeq \lambda T^2/2$$

indicating that delay increases as the square of the critical gap T.

If the critical gap is assumed to vary from individual to individual, the mean delay can be obtained by averaging out over T giving, approximately, $1/2\ \lambda\mu_2'$, where μ_2' is the second moment about the origin of the distribution of critical gap. This indicates that the delay is insensitive to the exact shape of the latter distribution, except perhaps for heavy traffic flows for which λ is large, but not to its mean and variance, since

$$\mu_2' \quad = \quad \text{variance} \ + \ (\text{mean})^2$$

The problem posed in Section 10.2 can now be solved. For $T = 8$ s, $\lambda = 1/6$ v.p.s., and the mean delay is expected to be $6[e^{4/3} - 7/3] = 8.8$ s. It should be noted, however, that we have been enabled to do more than answer a specific question. The remarks of the last paragraph indicate how even very simple models can be used to give a general insight into various aspects of the real situation under study.

10.4. More complicated models for intersections

The simple model considered in the last section can be made more realistic in a number of ways. For instance, we ought to consider, not a single vehicle on the minor road, but a queue of vehicles. If the flow along the major road is light, the assumption of random arrivals is sufficiently realistic. The major road flow can be taken as a general distribution or particularised as one of the alternatives to the Exponential suggested in Sections 10.2 or 10.5. Realistically, the major road flow should consist of at least two lanes in the same direction. (As indicated earlier, the problem of two lanes in opposite directions is mathematically intractable.)

If a queue of n $(n > 1)$ vehicles is waiting in the minor road, the following situation can be postulated. The leading driver considers a gap t and makes a decision whether or not to cross on some gap-acceptance criterion, which can be written generally as $\Gamma(t)$. Possible alternatives to the step-function already used are suggested in Section 10.6. If the leading driver decides to cross, the second vehicle moves up into the leading position and the procedure is repeated with the residual gap left after the first manoeuvre. Various models of this general type have been analysed theoretically, usually with some restrictions on the decision rules. For example, one such model incorporates the restriction that one vehicle, at most, crosses the major road in each gap of the major road flow. Entirely realistic models are difficult to describe theoretically and some such models have been criticised as being artificial. A critical survey of various types of model has been given by Breiman (1969). The most suitable method in practice would appear to be simulation, or perhaps a combination of theory and simulation.

10.5.   Headway distributions

The displaced Exponential has been known for some time to be a reasonably good model for free-flowing traffic with flow rates of up to about 800 v.p.h., but it breaks down for higher flows.   In urban conditions the presence of traffic lights and pedestrian crossings etc., causes bunching and a number of alternative distributions has been suggested (Buckley, 1962; Ashton, 1971).

Possible distributions suggest themselves in two ways.   In the first method, data collected on an open road can be plotted in a histogram, the general shape of which is usually similar to that of *Figure 10.5*. Visual inspection then suggests a likely theoretical formula.   In this

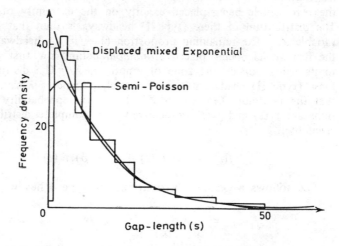

Figure 10.5   Histogram of headway data and fitted distributions

case a unimodal, positively skew distribution in the range $(0, \infty)$ is required.   The Gamma and Log normal distributions are roughly of this shape, although attempts to fit these distributions to data have not yielded very good results.

The second method involves starting from some hypothesis about how traffic moves down a road, in other words by postulating some mechanism governing the flow.   Any distribution thus arrived at must, of course, be validated subsequently by data-fitting procedures.   These procedures have been carried out for a number of distributions but there seems to be no real evidence to show that the displaced Exponential can be much improved upon and a satisfactory distributional form is still awaited.

Two examples of the latter method follow.   For the first model, suppose that a single lane of traffic can be divided roughly into two groups of vehicles.   Suppose that light, free-flowing traffic such as private cars form a proportion p of the total, and that the remaining $(1 - p)$ is made up of heavy goods vehicles.   If each group can be taken to proceed independently at random with associated flow-rates λ_1 and λ_2, respectively, the resultant Mixed Exponential distribution has probability density function

$$f(t) \;=\; p\lambda_1 e^{-\lambda_1(t-a_1)} + (1-p)\,\lambda_2 e^{-\lambda_2(t-a_2)},$$

$$t > \min(a_1, a_2), \quad 0 < p < 1$$

The second model is mathematically more complicated. It can be set up on the following assumptions. Suppose that behind each vehicle is a zone which vehicles never enter, called the *zone of emptiness*, and denoted Z, where Z is a random variable having a given distribution. A proportion p $(0 < p < 1)$ of headways result from the rear vehicle being placed exactly on the extremity of this zone. The distribution of these (type I) headways is then that of the random variable Z. The remaining proportion $(1-p)$ of headways result from the rear vehicle being placed *at random* behind the first one, subject to its being outside the zone of emptiness Z. The distribution of these (type II) headways is Exponential, parameter λ say, displaced so that the minimum headway is Z. If the two probability density functions are $f_1(t)$ and $f_2(t)$, respectively, the complete distribution is given by

$$f(t) \;=\; p\,f_1(t) + (1-p)\,f_2(t)$$

If Z follows a Gamma distribution, the type I headways have p.d.f.

$$f_1(z) \;=\; \frac{1}{\beta^\kappa\,\Gamma(\kappa)}\; z^{\kappa-1}\,e^{-z/\beta}\,, \quad z > 0$$

Let the p.d.f. for type II headways be $f_2(x)$: to determine $f_2(x)$. The variable X is distributed in a displaced Exponential for which $X \geqslant Z$. If $F_2(x)$ is the distribution function for X, we have

$$F_2(x) \;=\; P[X < x] \;=\; P[Y < x | Y > Z]$$

where Y is an Exponential variate with p.d.f. $g(y) = \lambda e^{-\lambda y}$, $y > 0$. Thus

$$F_2(x) \;\equiv\; \frac{P\,[Z < Y < x]}{P\,[Y > Z]}$$

Since the probability in the denominator is independent of x,

$$f_2(x) \;=\; \frac{d}{dx}\,F_2(x) \;=\; \frac{1}{P\,[Y > Z]}\;\frac{d}{dx}\,P[Z < Y < x]$$

But

$$P[Z < Y < x] = \int_0^x \int_0^y g(y) \, f_1(z) \, dz \, dy$$

thus

$$\frac{d}{dx} P[Z < Y < x] = \int_0^x g(x) \, f_1(z) \, dz$$

Substitution for $g(x)$ and $f_1(z)$ yields, for the right-hand side,

$$\frac{\lambda e^{-\lambda x} \, \gamma(\kappa, x/\beta)}{\Gamma(\kappa)}$$

where $\gamma(\kappa, x/\beta)$ is the incomplete gamma function. Also

$$P[Y > Z] = \int_0^\infty \int_0^y \lambda \, e^{-\lambda y} \, \frac{1}{\beta^\kappa \, \Gamma(\kappa)} z^{\kappa-1} \, e^{-z/\beta} \, dz \, dy, \quad z > 0$$

Inversion of the order of integration, followed by the substitution $y = z(1/\beta + \lambda)$, leads directly to the result

$$P[Y > Z] = \frac{1}{(1 + \lambda\beta)^\kappa}$$

so that

$$f_2(x) = \frac{\lambda e^{-\lambda x} \, \gamma(\kappa, \, x/\beta)(1 + \lambda\beta)^\kappa}{\Gamma(\kappa)}, \quad x > 0$$

Finally, writing t for z and x, we get

$$f(t) = p \, f_1(t) + (1 - p) \, f_2(t)$$

$$= \frac{p \, t^{\kappa-1} \, e^{-t/\beta}}{\beta^\kappa \, \Gamma(\kappa)} + \frac{(1 - p) \lambda e^{-\lambda t} \, \gamma(\kappa, \, x/\beta)(1 + \lambda\beta)^\kappa}{\Gamma(\kappa)}, \quad t > 0$$

$$(3)$$

A special case of $f(t)$ can be obtained by putting $\kappa = 1$ in (3): this corresponds to an Exponential zone of emptiness. Thus

$$f(t) = p \, \frac{e^{-t/\beta}}{\beta} + (1 - p) \lambda(1 + \lambda\beta) \, e^{-\lambda t} \, (1 + e^{-t/\beta}), \quad t > 0$$

The mixed Exponential and semi-Poisson distribution given by (3) are shown fitted to some data in *Figure 10.5*.

10.6. Gap-acceptance models

The model discussed in Section 10.4 incorporated the simplest possible gap-acceptance criterion, namely the step-function illustrated in *Figure 10.2*. Other possibilities are the trapezoidal function and the displaced Exponential, the distribution functions for which are illustrated in *Figure 10.6*.

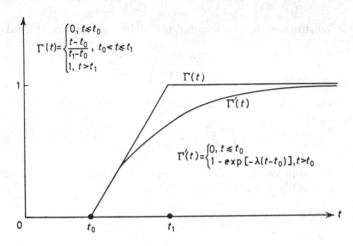

$$\Gamma(t)=\begin{cases}0, & t\leq t_0\\ \dfrac{t-t_0}{t_1-t_0}, & t_0<t\leq t_1\\ 1, & t>t_1\end{cases}$$

$$\Gamma'(t)=\begin{cases}0, & t\leq t_0\\ 1-\exp[-\lambda(t-t_0)], & t>t_0\end{cases}$$

Figure 10.6 Trapezoidal and displaced Exponential functions for gap-acceptance

It is also possible to use a general function in which every gap t has, associated with it, a probability $\gamma(t)$ of its being accepted.

10.7. Validation of the model

Validation of flow distributions or gap-acceptance functions forms a vital part of model-building. This procedure can cause problems. An indication of the kind of problems encountered in the validation of gap-acceptance functions is given here as an illustration. Difficulties in using data collected at a real intersection can be of both a conceptual and a practical kind. Ideally one would like to set up a situation in which vehicles in the main stream travel past the waiting vehicle, with the gaps gradually increasing in size at, say half-second intervals. In this way the first (i.e. minimum) gap accepted would be recorded. Strictly the experiment would need to be repeated a number of times before the minimum acceptable gap for any driver, or series of drivers, could be determined. This procedure is obviously impracticable, and data are usually collected by monitoring an actual junction and recording the sizes of gaps accepted and rejected by a series of drivers arriving at the GIVEWAY sign. Considerable care is then needed in the interpretation of any analysis based on this data.

Note first that the distribution of the total number of rejected gaps is made up of two components, corresponding to 'within a driver' and 'between drivers', respectively. This means that there is a mixture of two distributions. It is also the case that each driver in general rejects a series of gaps for each one that he accepts, thus for each driver and each accepted gap there is a *distribution* of rejected gaps, none of which is necessarily the 'maximum rejectable'. Even if the latter could be obtained, they would not, in general, be the same for each driver at different times. The distribution of accepted gaps is similarly anomalous, except that in this case only one gap is accepted per driver on any particular occasion. These gaps are not necessarily 'minimum acceptable'. In addition, since there are many more rejected than accepted gaps, the cautious driver is over-represented when all the data are used. Various methods for dealing with these difficulties have been devised, including the use of data on lags only, but this is wasteful of information.

10.8. References

ASHTON, W.D. (1966). *The Theory of Road Traffic Flow.* Methuen; London
ASHTON, W.D. (1971). 'Distributions for gaps in road traffic', *J. Inst. Math. Appl.,* **7**, 37
BREIMAN, L. (1969). 'Data and models in homogeneous, one-way traffic flow', *Transportation Research,* **3**, 235
BUCKLEY, D.J. (1962). 'Road traffic headway distributions', *Proc. First Conf. of Australian Road Research Board*
HOGG, R.V. and CRAIG, A.T. (1970). *Introduction to Mathematical Statistics.* Macmillan; London
LIGHTHILL, M.J. and WHITHAM, G.B. (1955). 'On kinematic waves II. A theory of traffic on long crowded roads'. *Proc. Roy. Soc. A,* **229**, 317

10.9. Problems for further study

1. Traffic flows along a one-lane road at random at a rate of 900 v.p.h. Taking a time-gap, in seconds, of $t = 1(1)15$ calculate

(a) The percentage of gaps which are longer than t.
(b) The percentage of time occupied by gaps longer than t.
(c) The mean of all time-gaps longer than t.

What differences in these calculations, if any, result from the use of a displaced Exponential incorporating a minimum headway of one second with the flow rate remaining unaltered.

2. The time-gaps, in seconds, between the instants at which vehicles pass a point on a road may be assumed to follow an Exponential distribution with probability density function $f(t) = \lambda e^{-\lambda t}$, $t > 0$. On a particular road the rate of flow of vehicles is 600 v.p.h. What is the mean length of gap between successive vehicles?

A pedestrian arrives just as a vehicle passes, and he crosses the road only when he judges there to be a gap of at least 4 s between successive vehicles. Assuming that errors of judgement can be ignored, calculate

(a) The probability that the pedestrian is delayed.
(b) The mean length of gap in which the pedestrian crosses.
(c) The mean length of time he waits, given that he must wait for just one vehicle to pass.
(d) The mean length of time he waits, if there is no restriction on the number of vehicles which pass whilst he does so.

3. An alteration scheme is proposed to a two-lane, one-way road with a view to aiding pedestrians in crossing. The proposal involves siting a refuge in the middle of the road. If the flows in the two lanes can be taken as independently random with rates λ_1 and λ_2, derive expressions for the probability that a pedestrian will be delayed and for the mean delay for each of the two situations before and after the alteration. Hence determine whether or not, on the criterion of delay to the pedestrian, the alteration is beneficial. What other criteria would it be sensible to consider?

4. Extend the simple model of Section 10.3 by making use of the gap-acceptance functions suggested in Section 10.6 and of different distributional forms, such as the Gamma or Mixed Exponential, for the major-road flow.

5. Bearing in mind that models of a stochastic nature break down for high rates of traffic flow, propose a suitable model for dense traffic.

11
A BUSINESS PLANNING MODEL

T. Lomas
*Management Services Department, Post Office Telecommunications Head-
quarters, London*

[Prerequisites: none]

11.1. Introduction

The Telecommunications Business of the Post Office is very large and
complex and the demands for its services are growing rapidly. Because
of the huge size and technological nature of the business the head-
quarters is organised into departments whose managers tend to specialise
in areas of activity, such as engineering development, technical planning,
marketing and service. In such an organisation it is difficult for managers always to see and judge actions in the context of the whole
business and its future. Thus business planning machinery has been
created so that each department's operational plan is related to an
overall business plan; the business plan broadly defining how the
business will achieve its principal objectives over the planning period
(in this case 11 years).

The business faces a decade of explosive growth during which the
system is expected to double in size. At the same time the underlying telecommunications technology is changing fast. The strategic
planning of such a large enterprise is complex and sensitive to quite
small variations in performance and in economic and financial trends.
To enable the Business to develop and maintain the most effective
strategic plans and ensure that it could properly meet its obligations,
it was necessary to design and commission a Telecommunications
Business Planning Model.

11.2. Major areas of the model

Figure 11.1 shows in outline the principles of operation of this model. Each different shaded area represents a sub-model which is here defined as part of the total model which can, if required, be used independently of the total model. These sub-models represent specialist planning processes as follows:

> INCOME
> CURRENT EXPENDITURE
> MANPOWER
> CAPITAL EXPENDITURE
> DEPRECIATION
> FINANCING

The model and its constituent sub-models have been developed and constructed as aids to business planning. They do not, in themselves, produce business plans but rather enable managers and planning staff to make sets of assumptions about future demands for services, pay rates, productivity, prices etc. and to see quickly the likely effects upon profit, return on capital, borrowing requirements, interest charges etc. The main purpose of the model is to enable senior managers to consider a range of possibilities and to choose the best plan of action. The model also allows for considerable experimentation and provides facilities for conducting sensitivity analyses whereby the sensitivity of the results to the assumptions can be explored.

Because of the large amount of data which is processed in the model (some 7000 pieces of input data simultaneously) the model is necessarily programmed on an off-line computer. Some of the sub-models are, however, also made available to on-line terminals for quick access. The following description of the Income sub-model illustrates how a typically simple sub-model is set up.

Most of the income to the business is derived from installation charges, rental charges, and call charges; the remainder is from Telex and miscellaneous services. For example if there are S_t installations in year t charged at u_t per installation then the installation income is $u_t S_t$. If there are T_{it} customers in rental class i in year t at r_{it} rent per customer then this part of the income is given by

$$\sum_{i=1}^{m} T_{it}\, r_{it}$$

and if there are D_{jt} calls made in call charge category j in year t at x_{jt} charge per calls then this part of the income is given by

$$\sum_{j=1}^{n} D_{jt}\, x_{jt}$$

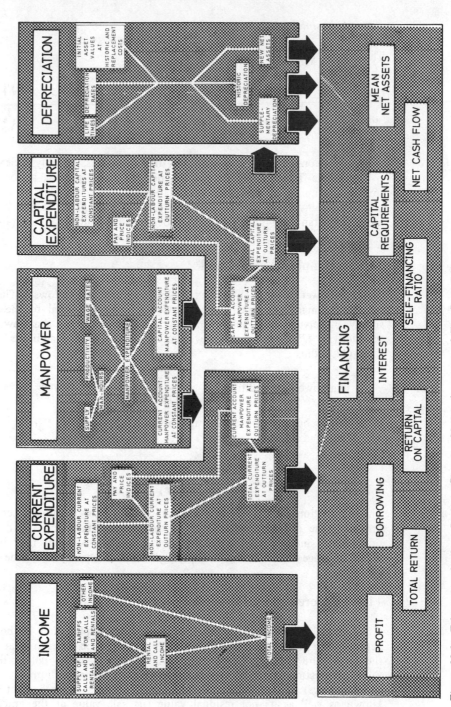

Figure 11.1 Telecommunications Business Planning Model (courtesy of The Post Office)

The total income for year t is thus

$$J_t \;=\; \underset{\substack{\uparrow \\ \text{total} \\ \text{income}}}{} u_t S_t \underset{\substack{\uparrow \\ \text{installations}}}{} + \sum_{i=1}^{m} T_{it}\, r_{it} \underset{\substack{\uparrow \\ \text{rents}}}{} + \sum_{j=1}^{n} D_{jt}\, x_{jt} \underset{\substack{\uparrow \\ \text{call} \\ \text{charges}}}{} + V_t \underset{\substack{\uparrow \\ \text{Telex} \\ \text{etc.}}}{} \quad (1)$$

where V_t represents the income from other services, notably Telex.

The whole Telecommunications Business Planning process is too large for its algebra to be described in detail here but some idea of the calculations involved may be deduced from *Figure 11.1*. Some of the more important sub-models are described in the following paragraphs.

11.3.   Manpower sub-model

The manpower sub-model, for example, is derived as follows. The number of man-hours spent on a service k is proportional to the supply of services (S_k) and inversely proportional to the manpower productivity (P_k). Now manpower expenditure $(E) \propto$ man-hours $\times$ wage rates (W_k) thus

$$\text{manpower expenditure} \;=\; \frac{\text{supply of services}}{\text{productivity}} \times \text{wage rates}$$

(each factor being expressed in appropriate units)

or

$$E \;=\; \sum_{k=1}^{n} \frac{S_k\, W_k}{P_k} \quad\quad (2)$$

The summation sign is to take account of the many different services provided, the different wage rates of the various skills and the different productivity indices relating to the various skills. Account may be taken of wage inflation by multiplying W by an appropriate index.

11.4.   Depreciation sub-model

A sub-model which perhaps needs some further explanation is the depreciation sub-model. Depreciation is a charge made each year on the profit and loss account of the business to cover the fall in the value of existing assets of the business (e.g. exchange equipment, cables and buildings) in that year. The expected lifetime of each asset is known as well as its net residual value (its sale value at the end of its life less the cost of its removal) and the depreciation is calculated to recover the original cost of the asset, less residual value, over its

estimated lifetime.   This depreciation is defined as historic depreciation.
Since during the lifetime of the assets the cost of replacing them will
probably increase, for example, because of pay and price increases, an
additional provision, called supplementary depreciation, is made to take
account of these increased costs.

The depreciation calculations can be summarised by three basic
equations.   These are:

Gross asset value at end of year = gross asset value at start of year
+ additions in year − value of
assets recovered in year

i.e.

$$G_{t+1} \;=\; G_t + A_t - W_t \tag{3}$$

Depreciation in year = gross value of assets at start of year
× depreciation rate

i.e.

$$D_t \;=\; G_t \times r \tag{4}$$

Net value of assets at end of year = net value of assets at start of
year + net capital expenditure
in year − depreciation in year

i.e.

$$N_{t+1} \;=\; N_t + C_t - D_t \tag{5}$$

The above calculations, applied separately to each asset group, are
repeated for each year of the planning period, the asset values being
up-dated each time.   This sub-model allows management to investigate
quickly the effects on depreciation of various investment strategies and
also the consequences of making changes in the depreciation rates and
rules, and in plant lives.

11.5.   Financing sub-model

It will be seen from the diagram of the Telecommunications Business
Planning Model that it is in the financing sub-model that the results
from the other sub-models are merged and some overall financial implications of, and indices for, the plan being evaluated are calculated.

This sub-model warrants detailed consideration for a number of
reasons.   Firstly, because it was, as an independent sub-model, immediately accepted and readily used by operational managers.   Secondly,
because it shows how, once a system has been described in mathematical

terms, the resultant model may be manipulated to answer questions which would have been extremely difficult, if not impossible, to answer before. Thirdly, because in its later stage of development it illustrates an application of mathematical programming to business planning. Finally, because it illustrates how a modular series of planning models might be developed catering for the requirements of both specialist managers and business planners.

While, as has already been said, the sub-model forms an integral part, in fact the principal part, of the total model it is also immediately available as required on a remote access time-sharing computer system. This on-line form is described here as it incorporates some facilities which were not required of the total model.

To appreciate the workings of the financing sub-model it is necessary to have some understanding of the financial aims of an enterprise such as the Post Office and some knowledge of how such an enterprise is financed.

The Post Office, as a nationalised industry, has no shareholders (other than government) requiring the payment of dividends. All profits are therefore returned to capital investment. It is required to satisfy government of its overall financial efficiency by meeting certain financial criteria, e.g. its income, year by year, must be sufficient to meet its outgoings and it must achieve an agreed ratio of return on capital. As it is in competition with other parts of the public sector for loans from central government the amount of money which it can borrow is limited and therefore it is required to finance a proportion of its investments from its own resources. In short it must be self-sufficient.

The financing sub-model enables planners to translate the implications of various alternative business plans in terms of income, expenditure, capital requirements etc. into effects upon interest, borrowing, profit, return on capital etc. The main input and output variables of this sub-model are shown in *Table 11.1*.

Table 11.1 Main variables used in the financing sub-model

INPUT		OUTPUT	
Variable	*Code*	*Variable*	*Code*
Income	J_t	Interest	I_t
Working expenditure	E_t	Borrowing	B_t
Capital requirements	C_t	Net assets	NA_t
Historic depreciation	DH_t	Profit	P_t
Supplementary depreciation	DS_t	Return on capital	ROC_t
Interest rate	i_t	Self financing ratio	SFR_t

Consider the inputs. The income from rentals, call charges and miscellaneous services charges has to cover to some extent both working expenditure and capital requirements of the business. Working expenditure includes the wages and salaries of operators, clerical staff, and general administration staff and the cost of all maintenance work. Working expenditure thus represents money flowing out of the business, i.e. a negative cash flow. The capital required by the business for investment in capital plant is provided partly from the business's own resources and partly from borrowing money from central government. The latter is, of course, paid for at an appropriate rate of interest.

Two other inputs to the financing sub-model are historic and supplementary depreciation on plant as described earlier.

In mathematical terms the condition that the sum of monies flowing into the business must be equal to that flowing out plus the increase in monies belonging to the business (current assets) is recognised as a continuity equation and expressed in the form

$$J_t + B_t = C_t + E_t + I_t \tag{6}$$

Loans from central government are at fixed interest rates. A number of such loans will be current in any one year and if we assume an average interest rate for any one year the total interest paid on loans in any year t is given by

$$I_t = \sum_{y=s}^{t-1} i_y B_y + f(B_t) \tag{7}$$

for loans from year s to year $t - 1$.

The term $f(B_t)$ refers to the fact that borrowing may start at any time during a year and not necessarily at the end; thus, interest is payable for part of the year in which the loan is initiated. For practical purposes it is acceptable to assume that $f(B_t)$ is directly proportional to the interest paid in a full year, thus

$$f(B_t) = a \, i_t \, B_t \tag{8}$$

From Equation (7) we can deduce that

$$I_t = I_{t-1} + a \, B_t \, i_t + (1 - a) \, B_{t-1} \, i_{t-1} \tag{9}$$

In practice a modified version of Equation (9) is used. B_t does not include additional loans which are created to cover the repayment of old loans. These latter do not affect the liability of the business but, as the new loans are made at different interest rates from the old loans, the interest payments in each year are increased (by q_t, say). Equation (9) thus becomes

$$I_t = I_{t-1} + a \, B_t \, i_t + (1 - a) \, B_{t-1} i_{t-1} + q_t \tag{10}$$

In the model q_t is calculated from input information on the loans due for renewal. The profit of the business in year t is written

$$P_t = J_t - E_t - DH_t - DS_t - I_t \tag{11}$$

An important factor to consider in measuring the financial performance of a business is the value of its assets. A critical measure of performance is the capacity of the assets to generate a return to the business (return on capital). Return on capital is generally defined as the ratio of return to asset value expressed as a percentage, or

$$\frac{\text{return}}{\text{asset value}} \times 100\% \tag{12}$$

In the Post Office Telecommunications Business it is more precisely defined as

$$\frac{\text{profits} + \text{interest charges} + \text{supplementary depreciation}}{\text{mean value of net assets throughout the year}} \times 100\% \tag{13}$$

Thus using (11) we obtain

$$ROC_t = \frac{J_t - E_t - DH_t}{(NA_t + NA_{t-1})/2} \times 100\% \tag{14}$$

where net assets, NA_t, are defined by Equation (5) which, in the notation of *Table 11.1*, is

$$NA_t = NA_{t-1} + C_t - DH_t$$

Another indicator which is of importance to business managers and business planners is the degree to which the business is financing its investments out of its own resources. This indicator is the self-financing ratio SFR and is defined as

$$SFR_t = \frac{P_t + DH_t + DS_t}{C_t} \times 100\% \tag{15}$$

Using (6) and (11) we have

$$SFR_t = \left(1 - \frac{B_t}{C_t}\right) \times 100\% \tag{16}$$

Equations (6) to (16) essentially describe the financing sub-model and, by their manipulation, the likely financial implications of alternative business strategies can rapidly be evaluated.

Suppose, for example, that all the inputs shown in *Table 11.1* have been provided and management wishes to examine the financing implications. The equations are applied year by year starting from the first year. Once the borrowing levels and interest charges for any year have been calculated the remaining equations can be solved. It will be seen that Equations (6) and (10) form a pair of simultaneous equations whose unknown factors are I_t and B_t. These equations can thus readily be solved.

It will be appreciated that the model is to be used not by mathematicians but by planners. The outputs from the model must, therefore, be in terms of real figures such as would be familiar to, for example, accountants and not in terms of mathematical formulae. *Table 11.2* shows a typical output from the financing sub-model.

It is most unlikely that the first plan to be evaluated by the model is satisfactory in all respects, satisfying simultaneously the various business targets and constraints, and so management will generally want to ask a number of questions. These will be likely to take the form of 'What must I change to achieve such and such a return on capital, or such and such a self-financing ratio, or to keep borrowing within such and such a limit?' Any one of these targets could be met or constraints satisfied by varying income, working expenditure, or capital requirements either singly or simultaneously. In practice income is allowed to vary while all the other inputs are held constant and one of the outputs is specified. This only approximately represents what is practicable but, in view of the long lead times on capital expenditure, it is sufficiently realistic to be acceptable for experimental purposes. Moreover it provides a useful estimate around which a new plan may be built.

With income allowed to vary Equations (6) and (10) form a pair of simultaneous equations with three unknowns one of which is specified by a third equation. Given a required return on capital, Equation (14) defines the required income. Given a target self-financing ratio, Equation (16) defines the amount to be borrowed. If the borrowing limit is constrained the calculations first proceed on the assumption that the initially forecast income will be sufficient to keep borrowing below its limit. If this is found not to be so then Equations (6) and (10) are solved for income and interest with borrowing set at the limit.

11.6. Use of the model in tariff decisions

A facility which is of particular interest to applied mathematics students is that provided for the evaluation of tariff policies, since it illustrates mathematical programming (specifically dynamic programming) in practice.

The basic aim of a tariff policy is to achieve a return on investments sufficient to meet the needs of the business whilst charging a fair price to its customers. With this aim constantly in mind it is, therefore, necessary from time to time to revise tariffs and, despite a consistently high record of improved productivity over the years it is occasionally necessary, in times of particularly high cost inflation, to revise these tariffs upwards.

Table 11.2 Typical output of financing sub-model. The figures used in this example are hypothetical and unconnected with any plans of the Post Office Corporation

---

WANT TO SEE INPUT SUMMARY? YES

SUMMARY OF INPUT

YR/E 31/3	INCOME (M.)	WORKING EXPTURE (M.)	CAPITAL REQMTS. (M.)	DEPRECIATION HIST. (M.)	SUPPY (M.)	MEAN NET ASSETS (M.)	INTRST RATE P.C.
1974	1100	700.0	700.0	180.0	75.0	3860	9.00
1975	1350	750.0	760.0	200.0	90.0	4400	9.00
1976	1500	800.0	800.0	220.0	105.0	4970	9.00
1977	1650	900.0	850.0	250.0	120.0	5560	9.00
1978	1900	1000.0	900.0	280.0	135.0	6170	9.00
1979	2100	1100.0	950.0	310.0	150.0	6800	9.00
1980	2300	1250.0	1000.0	340.0	165.0	7450	9.00
1981	2500	1400.0	1100.0	370.0	190.0	8145	9.00
1982	2800	1550.0	1200.0	400.0	205.0	8910	9.00
1983	3100	1700.0	1300.0	430.0	220.0	9745	9.00

YR/E 31/3	BRWING (M.)	PROFIT (M.)	INTRST (M.)	R.O.C. P.C.	S.F.R. P.C.	T.RETRN (M.)	N.C.F. (M.)
1974	490.1	−45.1	190.1	5.70	29.99	220.0	300.0
1975	392.6	77.4	232.6	9.09	48.34	400.0	160.0
1976	368.5	106.5	268.5	9.66	53.93	480.0	100.0
1977	404.1	75.9	304.1	8.99	52.46	500.0	100.0
1978	339.9	145.1	339.9	10.05	62.23	620.0	.0
1979	321.3	168.7	371.3	10.15	66.18	690.0	−50.0
1980	352.5	142.5	402.5	9.53	64.75	710.0	−50.0
1981	438.2	101.8	438.2	8.96	60.17	730.0	.0
1982	428.7	166.3	478.7	9.54	64.28	850.0	−50.0
1983	418.3	231.7	518.3	9.95	67.83	970.0	−100.0

WHICH TARGET + START YEAR TO SPECIFY? ROC, 1974
WHAT RETURN ON CAPITAL (PERCENT) SPECIFIED FOR EACH
YEAR? 11.0, 11.0, 11.0, 11.0, 11.0, 11.0, 11.0, 11.0, 11.0, 11.0,

WITH TARGET RETURN ON CAPITAL THE OUTCOME IS:-

YR/E 31/3	BRWING (M.)	PROFIT (M.)	INTRST (M.)	R.O.C. P.C.	S.F.R. P.C.	INCOME (M.)	CAP. RQ. (M.)
1974	279.1	165.9	183.7	11.00	60.13	1304.6	700.0
1975	286.4	183.6	210.4	11.00	62.31	1434.0	760.0
1976	270.4	204.6	237.1	11.00	66.21	1566.7	800.0
1977	250.5	229.5	262.1	11.00	70.52	1761.6	850.0
1978	226.6	258.4	285.3	11.00	74.82	1958.7	900.0
1979	198.2	291.8	306.2	11.00	79.13	2158.0	950.0
1980	164.9	330.1	324.4	11.00	83.51	2409.5	1000.0
1981	175.0	365.0	340.9	11.00	84.10	2666.0	1100.0
1982	178.0	417.0	358.1	11.00	85.17	2930.1	1200.0
1983	173.4	476.6	375.3	11.00	86.66	3202.0	1300.0

WHICH TARGET + START YEAR TO SPECIFY?   STOP

Moreover, it is not practicable to change tariffs when only small increases in income are required. A method was therefore needed to translate the additional income requirements into forecasts of when tariff increases should be planned and what the magnitudes of these increases should be.

A general specification of the problem will be given. For simplicity let us assume that all tariff changes occur at the beginning of a year. Let K_t be the additional income required in year t over that originally forecast. This is found by manipulation of the model as described earlier to meet, for example, a target return on capital. (In the remainder of this section K_t will be referred to as income.) Let T_t be the income from a change in tariff in year t and r_t the rate of growth of the business in the same year.

This latter is relevant because a change in tariff which initially produces an income of T_t per annum will in the next year produce an income of $r_{t+1} T_t$. In general, a tariff change in year t giving a change in income of T_t in that year will produce a change in income in year u of

$$T_t \, r_{t+1} \, r_{t+2} \, \cdots \, r_u \;=\; \frac{T_t \, R_u}{R_t}$$

where

$$R_t \;=\; \prod_{v=1}^{t} r_v$$

The change in income from a tariff change in any one year is restricted to fixed increment changes m with some minimum change km,

$$T_t \in \{0, \pm km, \pm (k+1)m, \ldots, \pm nm\}$$

where n and k are integers. The restriction on the interval between tariff changes is expressed by introducing zero-one dummy variables, δ_t such that

$$\delta_t = \begin{cases} 1 \text{ if there is a tariff change in year } t \\ 0 \text{ otherwise} \end{cases}$$

If tariff changes cannot occur less than M years apart, then we must also have

$$\sum_{u=0}^{M-1} \delta_{t+u} \leqslant 1, \quad \text{for all } t$$

The choice of optimising function is somewhat arbitrary. The tariff steps should be chosen to produce an income as near to the required income as possible. One method could be to minimise the squared

deviations from the required income. In this case the problem can be specified as:
Minimise

$$\sum_{t=1}^{N} \left(K_t - \sum_{u=1}^{t} \delta_u \ T_u \ \frac{R_t}{R_u} \right)^2 \tag{17}$$

subject to

$$\sum_{u=0}^{M-1} \delta_{t+u} \leqslant 1, \qquad \text{for } t = 1, 2, ..., N$$

where

$$T_t \in \{0, \pm km, \pm (k + 1) \ m, ..., \pm nm\}$$

The problem as specified above is similar to one which could be solved by dynamic programming.

Dynamic programming is a mathematical programming technique first developed by Bellman (1957), and since widely used in the operational research field. Mitchell (1972) provides an introduction to the method. Essentially the technique works by reducing the problem to a hierarchy of simple sub-problems each of which can be solved in order. Thus the tariff problem requires decisions on tariffs in each year t. If the best tariff policy to choose in year t depends on earlier decisions only in that they affect the total income at the start of year t, then the problem can be split into a hierarchy of problems of the form: 'What is the best policy to choose in year t, if the income at the start of year t is a?' If the best policy to choose in year $t + 1$ for any given income level a at the start of year $t + 1$ is known, then the problem reduces to calculating the effect of making a certain tariff change in year t and following the best policy in later years. Mathematically let $C(a, t)$ be the value of the optimum tariff policy with income a at the start of year t, that is $C(a, t)$ is the sum of squared differences for years $t, t + 1, ..., n$. Then, if tariff changes may occur one year apart, $C(a, t)$ is defined by

$$C(a, t) = \min_{\beta} \{ V_t \ (a, \beta) + C \ (\beta, t + 1) \} \tag{18}$$

where $V(a, \beta)$ = the squared difference in year t of an increase in income from a to β in that year = $(K_t - \beta)^2$. Equation (18) is solved for years $t = N, N - 1, ..., 2, 1$ in order and $(\beta - a)$ gives the optimum tariff in each year. The solution to the overall problem is given by $C(0, 1)$. A similar problem specification can be made if tariff changes can occur not less than M years apart.

The efficiency of the above method depends on the range of possible values which a may take in any year. Using the problem as specified in Equation (17), a can cover a large range of values as the size of income in year t depends on the year in which previous tariff increases were introduced. Under this situation, the dynamic programming method is equivalent to complete enumeration; the advantage comes if the possible values income can take in any year can be reduced to a small number of possibilities. This can be achieved by redefining the problem so that the possible tariff steps are inflated with time, and the optimising function is changed slightly. The effect of this is that the income in any year can now take only one of the values $m\lambda R_t$, where λ is a non-negative integer. Thus the problem is to minimise

$$\sum_{t=1}^{N} \frac{1}{R_t^2} \left(K_t - \sum_{u=1}^{t} \delta_u \, \phi_u \, R_t \right)^2 \tag{19}$$

subject to

$$\sum_{u=0}^{M-1} \delta_{t+u} \leqslant 1, \text{ for } t = 1, 2, ..., N$$

where $\phi_u = T_u/R_u \in \{0, \pm km, \pm (k+1) m, ..., \pm \lambda m, ..., \pm nm\}$

The problem can now be formulated as a dynamic programming problem. Let $C(\lambda, t)$ denote the value of the best policy in years $t, t+1, ..., N$ if the income at the start of year t is $m\lambda R_t$. The problem is now to find $C(0, 1)$, given that

$$C(\lambda, t) = \min \{ D(\lambda, t), E(\lambda, t) \} \tag{20}$$

where $D(\lambda, t)$ = value of having no tariff change in year t

$$= \left(\frac{K_t}{R_t} - \lambda m \right)^2 + C(\lambda, t+1)$$

and $E(\lambda, t)$ = value of having a tariff change in year t

$$= \min \left\{ \sum_{u=0}^{M-1} \left(\frac{K_{t+u}}{R_{t+u}} - \lambda'm \right)^2 + C(\lambda', t+M) \right\}$$

$$\text{for } k \leqslant |\lambda' - \lambda| \leqslant n$$

the equations being solved in order for years $N, N-1, ..., 2, 1$.

Once the problem as defined in Equation (19) has been solved, a near optimum solution to Expression (17) can be found by rounding the tariff steps.

The dynamic programming approach is particularly flexible for alternative problems. A number of additional facilities have been incorporated in the model including the facility to consider tariff changes at times other than at the beginning of the year, tariff changes at fixed intervals and tariff changes at specified dates.

11.7.   Example of tariff calculations

Suppose that a provisional plan has been set up and analysed using the financing sub-model.   A comparison of the results against targets showed that to obtain the desired result would require additional income K_t in each year as shown in *Table 11.3.*

Table 11.3   Additional income (£ millions)

Year:	1	2	3	4	5	6	7	8	9	10
K_t	10	55	92	121	140	120	112	132	184	258
K_t/R_t	10	50	80	100	110	90	80	90	120	160

Assume the following constraints apply to tariff packages.   Tariff changes can occur not less than 2 years apart; tariff changes will increase income by steps of £10million and each tariff change must result in a change in income of at least £20million.   The system is growing by 4.9% per annum.   In other words $r_t = \sqrt{1.1}$, $k = 2$, $m = 10$ and $M = 2$ in the notation of Equations (19) and (20).

The problem is solved by progressively calculating all the values $C(\lambda, t)$, where t takes the values 1, 2, ..., 10 and λ the values 0, 10, 20, ..., 170.   These values are shown in *Table 11.4* and are calculated as described in Equation (20).   The income λ' which gives the best policy is shown in brackets below the value.   If the best policy is not to have a tariff change (i.e. if $\lambda' = \lambda$) the figure is marked with an asterisk.

The columns of *Table 11.4* are calculated in reverse order starting with year 10.   The figures in this column give the minimum squared error in year 10, provided a tariff change is possible in that year, for each possible income value at the end of year 9.   Thus, it is required to minimise $1.1^{-10} (160 - 10\lambda')^2$, and it will be seen that this can be done by setting $\lambda' = 16$ and hence $C(\lambda, 10) = 0$, except where $\lambda = 15$ or 17.   In these two cases λ' cannot be 16 as this would imply an income change of only £10million; the best policy here is to get as near to 160 as possible, i.e. to have no tariff change.

Column 9 of *Table 11.4* may now be calculated.   Here two situations must be distinguished.   If a tariff change is made in year 9 the new income level λ' must be the same for year 10.   The squared error is

$$1.1^{-9} (120 - 10\lambda')^2 + 1.1^{-10} (160 - 10\lambda')^2$$

and the figure λ' which minimises this value (subject to $|\lambda' - \lambda| \geqslant 2$) is required.   For example if $\lambda = 7$ then

157

Table 11.4 Calculation of optimum tariff steps. The table shows the calculated values of $C(\lambda, T)$. Figures in brackets give the best policy to follow

Income (£10 million)	Year 1	Year 2	Year 3	Year 4	Year 5	Year 6	Year 7	Year 8	Year 9	Year 10
0	809 (0)*	718 (6)	608 (8)	335 (10)	489 (10)	273 (8)	370 (8)	216 (10)	324 (14)	0 (16)
1	718 (1)*	718 (6)	608 (8)	335 (10)	489 (10)	273 (8)	370 (8)	216 (10)	324 (14)	0 (16)
2	809 (2)*	718 (6)	608 (8)	335 (10)	489 (10)	273 (8)	370 (8)	216 (10)	324 (14)	0 (16)
3	1082 (3)*	718 (6)	608 (8)	335 (10)	489 (10)	273 (8)	370 (8)	216 (10)	324 (14)	0 (16)
4	1443 (2)	691 (4)*	608 (8)	335 (10)	489 (10)	273 (8)	370 (8)	216 (10)	324 (14)	0 (16)
5	1302 (3)	608 (5)*	608 (8)	335 (10)	489 (10)	273 (8)	370 (8)	216 (10)	324 (14)	0 (16)
6	1302 (3)	691 (6)*	608 (8)	335 (10)	489 (10)	273 (8)	370 (8)	216 (10)	324 (14)	0 (16)
7	1302 (3)	741 (7)*	410 (7)*	335 (10)	386 (10)	280 (9)	268 (7)*	216 (10)	324 (14)	0 (16)
8	1302 (3)	718 (6)	335 (8)*	335 (10)	335 (10)	273 (8)*	216 (8)*	216 (10)	324 (14)	0 (16)
9	1302 (3)	718 (6)	416 (9)*	341 (11)	529 (9)*	280 (9)*	280 (9)*	229 (11)	324 (14)	0 (16)
10	1302 (3)	718 (6)	609 (8)	335 (10)*	335 (10)*	273 (8)	370 (8)	216 (10)	170 (10)*	0 (16)
11	1302 (3)	718 (6)	608 (8)	341 (11)*	273 (11)*	273 (8)	370 (8)	229 (11)*	42 (11)*	0 (16)
12	1302 (3)	718 (6)	608 (8)	335 (10)	335 (12)*	273 (8)	370 (8)	216 (10)	0 (12)*	0 (16)
13	1302 (3)	718 (6)	608 (8)	335 (10)	489 (10)	273 (8)	370 (8)	216 (10)	42 (13)*	0 (16)
14	1302 (3)	718 (6)	608 (8)	335 (10)	489 (10)	273 (8)	370 (8)	216 (10)	324 (14)*	0 (16)
15	1302 (3)	718 (6)	608 (8)	335 (10)	489 (10)	273 (8)	370 (8)	216 (10)	389 (13)	39 (15)*
16	1302 (3)	718 (6)	608 (8)	335 (10)	489 (10)	273 (8)	370 (8)	216 (10)	324 (14)	0 (16)*
17	1302 (3)	718 (6)	608 (8)	335 (10)	489 (10)	273 (8)	370 (8)	216 (10)	324 (14)	39 (17)*

$\lambda' = 5$ gives error $= 1.1^{-9} (120 - 50)^2 + 1.1^{-10} (160 - 50)^2 = 6743$
$\lambda' = 9$ gives error $= 1.1^{-9} (120 - 90)^2 + 1.1^{-10} (160 - 90)^2 = 2271$
$\lambda' = 10$ gives error $= 1.1^{-9} (120 - 100)^2 + 1.1^{-10} (160 - 100)^2 = 1557$

....

$\lambda' = 13$ gives error $= 389$

$\lambda' = 14$ gives error $= 324$

$\lambda' = 15$ gives error $= 420$

$\lambda' = 16$ gives error $= 679$

Consideration of the other possibilities shows that the minimum squared error occurs where $\lambda' = 14$ and the error is then 324. (Thus, in the notation of Equation (20), $E(7, 9) = 324$.) The second situation is to leave income unchanged and to follow the best policy from the next year, i.e. $D(7, 9)$ is required. By definition

$$D(7, 9) = 1.1^{-9} (120 - 70)^2 + C(7, 10) = 1060 + 0 = 1060$$

This is the squared error in having no tariff change in year 9 and an increase to income of £160million in year 10. In this case $E(7, 9) < D(7, 9)$ and the optimum tariff change is given by $\lambda' = 14$.

The remaining columns of the table are found in a straightforward manner using Equation (20). It may be helpful to appreciate that $C(\lambda, t)$ gives the least future squared error for all later years *provided that it is possible to have a tariff change in year* t. Thus, if a tariff change is made in year t the squared errors in year t and $t + 1$ are fixed and the best policy is then followed in year $t + 2$. Otherwise the squared error in year t is known and the best policy can be followed from year $t + 1$. For example if $\lambda = 8$ and $t = 5$, $E(8, 5)$ is found by the following calculations:

$\lambda' = 5$; error $= 1.1^{-5} (110 - 50)^2 + 1.1^{-6} (90 - 50)^2 + C(5, 7)$
$= 2235 + 903 + 370 = 3508$

$\lambda' = 6$; error $= 1.1^{-5} (110 - 60)^2 + 1.1^{-6} (90 - 60)^2 + C(6, 7)$
$= 1552 + 508 + 370 = 2430$

$\lambda' = 10$; error $= 1.1^{-5} (110 - 100)^2 + 1.1^{-6} (90 - 100)^2 + C(10, 7)$
$= 62 + 56 + 370 = 489$

$\lambda' = 11$; error $= 1.1^{-5} (110 - 110)^2 + 1.1^{-6} (90 - 110)^2 + C(11, 7)$
$= 0 + 226 + 370 = 596$

$\lambda' = 12$; error $= 1.1^{-5} (110 - 120)^2 + 1.1^{-6} (90 - 120)^2 + C(12, 7)$
$= 62 + 506 + 370 = 938$

and thus $E(8, 5) = 489$ with an optimum income of £100million. Considering the possibility of no tariff change

$$D(8, 5) = 1.1^{-5} (110 - 80)^2 + C(8, 6) = 832$$

and the best solution is to change to an income level of $\lambda' = 10$, i.e. £100million.

Once the table has been completed the minimum squared error is given by $C(0, 1)$. The various income levels, and hence tariff changes may also be found from *Table 11.4.* as follows. Starting with $C(0, 1)$ look at the best income shown. If this implies a tariff change (income level $\lambda' \neq \lambda$) then go to $C(\lambda', t + 2)$ and look at the best income in that position. Otherwise go to $C(\lambda, t + 1)$ and look at the best income there. In this way the optimum income level can be followed through the table. The results of this procedure are shown in *Table 11.5.* The tariff changes may now be inflated by the system growth (for example 10% per annum) and rounded to the nearest (say) £10million.

Table 11.5 Solution of the tariff problems

Year:	1	2	3	4	5	6	7	8	9	10
K_t/R_t	0	60	60	100	100	80	80	100	100	160
Tariff	0	60	0	40	0	−20	0	20	0	60
Rounded tariff	0	70	0	50	0	−30	0	30	0	100

11.8. References

BELLMAN, R. (1957). *Dynamic Programming,* Princeton University Press; Princeton, New Jersey

GOCH, D. (1969). *Finance and Accounts for Managers.* Pan Books Ltd.; London

MITCHELL, G.H. (1972). *Operational Research: Techniques and Examples,* English University Press; London

11.9. Problems for further study

1. What are the main purposes of a business planning model? How would you use such a model to investigate the relative importance of the various data required by a business in its planning function?

2. Equation (1) defines the total income to the business in terms of installations, rents, call charges, Telex and other services as though these incomes were independent. In practice the number of calls made is dependent upon, among other things, the number of telephones. Show how Equation (1) might be modified to take account of this interrelationship.

3. The financing sub-model is appropriate, as described, to a national-ised industry. How might this sub-model be modified so that it might be used in a similar situation in the private sector?

4. The example in the text on the calculation of a tariff policy assumed that each tariff package must increase income by at least

£10million.   How would the calculations be changed if this limit was raised to £40million? Describe, without necessarily completing all the calculations, how you would calculate a required policy.

5.   It is required to improve the return on capital ratio of the business by increasing tariffs and thus raising additional income. Assuming that working expenditure, capital requirements, and depreciation are unchanged, devise an expression for the change in self-financing ratio resulting from a given change in return on capital ratio.   Under what circumstances do you think the above assumptions are valid?

6.   The inputs to the financing sub-model are not independent.   For example, depreciation depends to some extent upon capital expenditure. By considering a company manufacturing a single commodity, develop mathematical formulae relating these inputs ultimately to the production level of the company.   Show how you would solve these equations to determine the level of production required in order to produce a return on capital of 15%.   What problems would be introduced if the company manufactured more than one product? What would your approach now be if the company manufactured two products?

7.   You are the head of a planning department responsible for forecasting income, running costs, and capital expenditures for your business over the next ten years.   How would you tackle this task? How might mathematical models help in this work? (*see* Problem (6) above).

12
THE CONTROL OF THE GRADE STRUCTURE IN A UNIVERSITY

D.J. Bartholomew
Department of Statistics, London School of Economics and Political Sciences

[Prerequisites: probability theory and Markov chains]

12.1. The problem

The following problem, which arose at an American university, is typical of those faced by many organisations at the end of a period of growth. The teaching staff were divided into three grades: full professor, associate professor and assistant professor. Although the total number of staff had ceased to increase, the number of senior staff continued to grow relative to the more junior. This constituted a problem— not because senior staff are undesirable — but because they cost more in salaries. At a time of budgetary standstill the prospect of a steadily rising salary bill posed the following two questions for administrators. Is the growth at the top likely to continue and, if so, what can be done to arrest or, if possible, reverse it?

Our object in this chapter is to consider how to formulate this problem in mathematical terms and then to contribute to its solution by mathematical analysis. Put in other terms, we shall aim to construct a mathematical model of the manpower system which we can then bring to bear on the questions raised above. The model will be constructed in two stages; first we shall give a quantitative description of the system and then make a set of assumptions about how changes take place.

12.2.   Stocks and flows

The central set of quantities in our problem is the numbers of people in each grade at a given time – the *stocks*. We shall use the notation $n_i(T)$ (i = 1, 2, ..., k), to denote the number of persons in grade i at time T. (There is no need at this stage to suppose that the grades are ranked according to seniority.) The stock numbers may change at any time but in the academic world most changes take place at the end of the academic year. We shall therefore approximate the behaviour of the system by assuming that changes take place only at annual intervals. T is thus expressed in years and is restricted to integer values.

Stock numbers change as a result of *flows*, both in and out of the system (recruitment and wastage) and by internal movement (mainly promotion). Suppose that out of the stock $n_i(T)$, $n_{ij}(T)$ people move to grade j by time $T + 1$ and that $n_{i,k+1}(T)$ leave the university altogether. The stock in grade i at time $T + 1$ thus consists of the survivors from time T plus the new entrants; the latter are denoted by $n_{0i}(T + 1)$. The relationship between the stocks and flows thus defined is

$$n_j(T + 1) = n_j(T) + n_{0j}(T + 1) + \sum_{\substack{i=1 \\ i \neq j}}^{k} n_{ij}(T) - n_{j,k+1}(T) - \sum_{\substack{i=1 \\ i \neq j}}^{k} n_{ji}(T)$$

$$= \sum_{i=1}^{k} n_{ij}(T) + n_{0j}(T + 1), \quad (j = 1, 2, ..., k) \tag{1}$$

if we define

$$n_{jj}(T) = n_j(T) - \sum_{\substack{i=1 \\ i \neq j}}^{k} n_{ji}(T) - n_{j,k+1}(T)$$

to mean the number of survivors in grade j.

These equations are simply accounting relations which, of themselves tell us very little. Their role is to make explicit the basic restraints under which the system operates. They also focus attention on what remains to be specified before the model is complete. The flows bring about changes in the stocks and we must therefore go on to make assumptions about how movements are generated. If we could predict the flows by some means we could clearly deduce the stocks in year $T + 1$ from those in year T and so on as far ahead as desired.

12.3.   Assumptions about flows

In a model we aim to reproduce as far as possible the characteristics of the real system which it is supposed to represent. At this stage we must therefore turn to data about the behaviour of the actual system to see what assumptions could be justified. The basis of all scientific forecasting is the identification of regularities in the past coupled with the assumption that they will persist into the future. Further progress is only possible after statistical investigation of past figures of stocks and flows. (We shall qualify this statement later.)

Promotion flows, with which we shall be primarily concerned, are governed by a variety of factors which vary from one type of employment to another. Sometimes the number of promotions is directly related to the number of vacancies occurring higher up; sometimes promotion is almost automatic on the attainment of a certain level of competence. In the particular university application mentioned at the beginning, the latter description was more realistic and so it was reasonable to look for a relationship between flow numbers and the stocks from which they originated. It turned out that the relationship was one of simple proportionality. That is, the ratios $n_{ij}(T)/n_i(T)$ $(i = 1, 2, ..., k + 1)$ were, apart from statistical fluctuations, constant over time. In practice it is quite common to find such a relationship even when the mode of operation of the system suggests it might be otherwise. However, this is something which must always be checked in practice so that alternative assumptions can be made if the evidence requires it.

We could now proceed to predict stock sizes on the assumption that $n_{ij}(T) \propto n_i(T)$ using an estimate of the constant of proportionality derived from our data. In adopting this course we should be treating the model *deterministically*. This would, in fact, be sufficient for the immediate purposes of this chapter but it would be unrealistic and misleading to take it too far. Although the ratios $n_{ij}(T)/n_i(T)$ may not depend on T in a systematic way they will certainly vary. This variation could be quite large if $n_i(T)$ is small, since leaving, in particular, is a highly unpredictable phenomenon at the individual level. A realistic model ought therefore to incorporate not only the regularities observed in the aggregate but the uncertainties of individual behaviour. Probability theory is the branch of mathematics which enables us to quantify uncertainty and hence we shall introduce a probabilistic (or stochastic) element into the model. Let us assume that movements take place independently and that an individual in grade i has probability p_{ij} of moving to grade j between one year and the next. Let his chance of leaving be w_i then, clearly,

$$\sum_{j=1}^{k} p_{ij} + w_i = 1 \qquad (i = 1, 2, ..., k) \qquad (2)$$

since an individual must stay where he is, move elsewhere or leave. Under this assumption the number flowing from i to j in a year will be a binomial random variable given the initial stock $n_i(T)$. Hence the *expected* flow will be $n_i(T) p_{ij}$ which accords with the assumed empirical observation that flows are proportional to stocks.

The remaining aspect of the system is recruitment. This can be conveniently looked at in two parts. First the total number recruited to the system and secondly the way they are distributed among the grades. In an organisation whose total size is fixed, as in the example at the beginning of this chapter, the total recruitment must be equal to the total loss. That is

$$R(T + 1) = \sum_{i=1}^{k} n_{i,k+1}(T) \qquad (3)$$

The way in which recruits are distributed between grades is often fairly constant being determined by the needs or policy of the organisation. Let us therefore assume that a proportion r_i is allocated to grade i ($i = 1, 2, ..., k$) with

$$\sum_{i=1}^{k} r_i = 1$$

Collecting these assumptions together, our model is thus specified by:

1. A matrix of transition probabilities governing movements within the system which we denote by $P = \{p_{ij}\}$.
2. A vector of wastage probabilities $w = (w_1, w_2, ..., w_k)$ related to the p_{ij}'s by Equation (2).
3. A vector of recruitment proportions $r = (r_1, r_2, ..., r_k)$.
4. A constraint,

$$\sum_{i=1}^{k} r_i = 1$$

12.4. The basic prediction equation

According to our model, next year's stocks are random variables and their values cannot, therefore, be predicted with certainty. In such circumstances we commonly use the *expected value* of the random variable as a predictor. (We ought to accompany such a prediction by a standard error and this is where the stochastic aspect of the model comes in but there is no space to develop this point here.)

Let us take expectations on both sides of Equation (1), conditional on the stocks at T. We have already noted that

$$\bar{n}_{ij}(T) = n_i(T) p_{ij}$$

where the bar over the n denotes the expectation. The recruitment to grade j, $n_{0j}(T + 1)$, may be written as $R(T + 1) r_j$ so we require the expectation of $R(T + 1)$. Now

$$\bar{n}_{i,k+1}(T) = n_i(T) w_i$$

so from (3)

$$\bar{R}(T + 1) = \sum_{i=1}^{k} n_i (T) w_i$$

Hence, on substituting in (1)

$$\bar{n}_j (T + 1) = \sum_{i=1}^{k} n_i (T) p_{ij} + r_j \sum_{i=1}^{k} n_i (T) w_i, \tag{4}$$

$$(j = 1, 2, ..., k)$$

These equations can be simply expressed in matrix notation as, say

$$\bar{n}(T + 1) = n(T) \{P + w'r\} = n(T) Q \tag{5}$$

Thus if the parameters of the model can be estimated, next year's (i.e. $T + 1$) stock can be predicted from this year's (year T) by a simple matrix multiplication. Next year's prediction, $\bar{n}(T + 1)$, can then be used as the base for prediction one further year ahead using

$$\bar{n}(T + 2) = \bar{n}(T + 1) Q \tag{6}$$

(We cannot put $n(T + 1)$ on the right-hand side because it is not known in year T; we therefore use its predicted value.)

The matrix Q is a special kind of matrix called a stochastic matrix and represents all possible grade-to-grade transitions. It has non-negative elements and all of its row sums are unity (*see* Problem (1)). Such matrices are central to the theory of Markov chains and so we can use that theory to answer questions about the behaviour of the model.

12.5. Prediction

The first question we raised about the university grade structure was whether the growth at the top was likely to continue. This question could be answered by using (6). Suppose that the initial stocks and the parameter values were as follows:

$$n(0) = (300, 100, 50)$$
$$w = (0.2, 0.1, 0.2)$$
$$r = (0.75, 0.25, 0)$$

and

$$P = \begin{bmatrix} 0.6 & 0.2 & 0 \\ 0 & 0.7 & 0.2 \\ 0 & 0 & 0.8 \end{bmatrix}$$

where the grades are listed in order of increasing seniority. The form of **P** given above is fairly typical. The zeros below the diagonal mean that there are no demotions; there are promotions into the next higher grade only. The wastage vector shows a high rate of loss at the top and bottom; the former would include death and retirement to a greater extent than in the two lower grades. Most recruitment is into the bottom grade.

The reader should construct the matrix **Q** and project the grade structure for 5 or 10 years ahead (*see* Problem (2)). The calculations will show a steadily deteriorating position as the system becomes increasingly top-heavy. This behaviour depends, of course, on the structure of **P** but we have taken a fairly typical case which, if anything, offers more favourable promotion prospects than occur in many organisations. The conclusion must be that the recruitment and promotion policies represented by **r** and **P** are incompatible with maintaining a structure like **n**(0). (The reader will notice that the elements of the vector of predicted grade sizes will not, in general, be integers; this is because we are dealing with expectations. The mathematician knows that expectations of integer valued random variables need not be whole numbers but in presenting results to management talk of fractional men sometimes tends to undermine confidence!)

Having forecast that things will get worse we ought to know just how bad they can get. In mathematical terms — what is the limiting behaviour of **n**(*T*) as $T \to \infty$? After *T* time periods

$$\bar{n}(T) \;=\; n(0)\,Q^T \tag{7}$$

The theory of Markov chains shows, under very general conditions which will be satisfied in any meaningful manpower problem, that

$$\lim_{T \to \infty} Q^T \;=\; Q^\infty \tag{8}$$

where Q^∞ is a stochastic matrix with identical rows. If **q** denotes this common row then, letting $T \to \infty$ in (7)

$$n(\infty) \;=\; n(0)\,Q^\infty \;=\; N\,q \tag{9}$$

where *N* is the total (fixed) size of the system. There is, therefore, a limiting structure which *does not depend on the starting structure*. The easiest way to calculate **q** is by observing that the limiting structure must satisfy

$$n(\infty) \;=\; n(\infty)\,Q \quad \text{or} \quad q \;=\; q\,Q \tag{10}$$

This system of equations is singular but if we omit one of them and use the fact that

$$\sum_{i=1}^{k} n_i(\infty) \;=\; N\!\left(\sum q_i \;=\; 1\right)$$

they can easily be solved. The reader should make this calculation and compare it with the succession of stock vectors (*see* Problem (3)).

12.6. Control: maintainability

Having discovered the inevitability of growth at the top with current rates the next problem is to control the situation. Let us take first the limited objective of trying to stay where we are. If *n* is the present structure which we wish to maintain then it must obviously satisfy

$$n = n Q \tag{11}$$

In mathematical terms the control problem is to find a matrix Q such that (11) is satisfied. However, Q is a function of P, w and r and these may not all be susceptible to control. Natural wastage, for example, is not under direct management control and dismissal is something which most employers would prefer to avoid. Promotion is under direct management control but lack of suitable promotees or a policy to fill vacancies by promotion might only allow promotion rates to be varied within narrow limits. The recruitment vector is also subject to direct control but here again limits might be set by the availability of qualified candidates or policy restraints.

The mathematical problem, with which we are faced is thus to find a Q, satisfying (11), subject to all the restraints which practical politics imposes on the operation of the system. It may, of course, be impossible to find a suitable policy at all.

To illustrate the solution we shall make a rather simple assumption which is, nevertheless, often realistic. Assume that P, and hence w, cannot be changed at all. All control must therefore be exercised through r which we suppose can be varied at will subject to

$$r \geqslant 0 \quad \text{and} \quad \sum_{i=1}^{k} r_i = 1 \tag{12}$$

(An inequality linking two vectors is to be understood as holding between each pair of elements.) In this case our problem can be solved by finding an r which satisfies (11) and (12). Noting that $Q = P + w'r$ it easily follows that

$$r = n(I - P)(n \ w')^{-1} \tag{13}$$

where I is the unit matrix; note that $n \ w'$ is a scalar. The reader should verify that the elements of r given by (13) sum to one. However, they will only all be non-negative if

$$n \geqslant n P \tag{14}$$

Thus we can easily check whether a particular structure is maintainable by recruitment control.

Such an arithmetical check serves the immediate purpose but it fails to give very much insight into what kind of structures are maintainable. We therefore go on to try to characterise the set of structures which satisfy (14).

Since the total system size is fixed let us work in terms of the proportions in each grade and denote them by $x = n\,N^{-1}$. We are thus interested in the set of x's satisfying

$$x \geqslant x\,P \qquad\qquad (15)$$

when $k = 3$ we can visualise the problem geometrically. The vector x can be represented as a point in 3-dimensional Euclidean space. Such an x must lie on the plane $x_1 + x_2 + x_3 = 1$ and lie in the positive octant. Hence the set of all possible structures can be represented by the set of all points in the equilateral triangle with vertices (1, 0, 0), (0, 1, 0) and (0, 0, 1) illustrated in *Figure 12.1*.

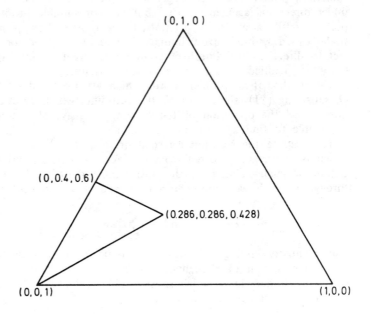

Figure 12.1   The maintainable region for recruitment control with k = 3 *and*

$$\mathbf{P} = \begin{bmatrix} 0.5 & 0.4 & 0 \\ 0 & 0.6 & 0.3 \\ 0 & 0 & 0.8 \end{bmatrix}$$

The Inequality (15) defines a region in this triangle containing all maintainable structures. If we could find the boundary of this region we would be able to see immediately what kind of structures are maintainable. We shall now do this algebraically by expressing any x satisfying (15) as a linear combination (linear function with positive coefficients summing to one) of a fixed set of vertices and so deduce that the maintainable region is the convex hull determined by these vertices. We shall work in terms of a general k but the geometrical terminology used for $k = 3$ will be retained.

Expressing (13) in terms of x,

$$x = xw'r (I - P)^{-1} \qquad (16)$$

Post multiplying both sides of (16) by a column vector of 1's, denoted by I' we find that

$$1 = xw'r (I - P)^{-1} I' = xw'rd' \qquad (17)$$

where the elements of d are the row sums of $(I - P)^{-1}$. Hence, substituting from (17) in (16)

$$x = (r\,d')^{-1}\, r\, (I - P)^{-1}$$
$$= \sum_{i=1}^{k} r_i\, e_i\, (I - P)^{-1} \Big/ \sum_{i=1}^{k} r_i\, d_i \qquad (18)$$

where e_i is a vector with 1 in the ith position and zeros elsewhere. Let

$$a_i = r_i\, d_i \Big/ \sum_{i=1}^{k} r_i\, d_i$$

then x may be written

$$x = \sum_{i=1}^{k} a_i\, \{e_i\, (I - P)^{-1}\, d_i^{-1}\} \qquad (19)$$

The coefficients a_i are non-negative and they sum to one. Any such point x thus lies in the convex region with vertices having co-ordinates

$$e_i\, (I - P)^{-1}\, d_i^{-1} \qquad (i = 1, 2, \ldots, k)$$

and each such point corresponds to a different r.

To illustrate the calculations suppose that

$$P = \begin{bmatrix} 0.5 & 0.4 & 0 \\ 0 & 0.6 & 0.3 \\ 0 & 0 & 0.8 \end{bmatrix} \quad \text{for which } (I - P)^{-1} = \begin{bmatrix} 2 & 2 & 3 \\ 0 & 2.5 & 3.75 \\ 0 & 0 & 5 \end{bmatrix}$$

Dividing each row by the row sum the vertices of the maintainable region are

$$(0, 0, 1), \quad (0, 0.4, 0.6) \quad \text{and} \quad (0.286, 0.286, 0.428)$$

These points are plotted on the figure and the maintainable region is a triangle containing top-heavy structures only. In fact, the least top-heavy structure is the vertex $(0.286, 0.286, 0.428)$ for which the top grade is twice the size of the two below it. This is typical of what one finds with upper triangular P's with large elements on the main diagonal. The reader should carry out the calculations for the numerical example in Section 12.5 (*see* Problem (4)).

A similar analysis can be made when only the promotion rates can be controlled. In this case we fix r and w and examine the effect of varying the elements of P subject to the restriction that $d_i = 1 - w_i$ for all i. For a general P the problem is complicated by the fact that there are infinitely many P's satisfying (11). However, if we are considering a simple hierarchy in which promotion is into the next higher grade only, P has non-zero elements only on the main diagonal and the superdiagonal. In this case there is a unique solution of (11) and the set of n's for which there is a P with non-negative elements is the maintainable region. Unlike the region for recruitment control this region turns out to include bottom-heavy structures. This result suggests that the maintenance of a bottom-heavy structure is likely to be tackled more successfully by control of promotion than by recruitment.

12.7.   Control: attainability

Analyses such as those carried out above give considerable insight into the circumstances under which a given structure can be maintained. They do not tell us how to reach a desired structure but only how to remain there when we do reach it. The problem of attainability is that of how to move from a given structure $x(0)$, say, to a desired structure x^*. A theory of attainability must tell us whether x^* can be attained and, if so, how. These are difficult questions and a full solution requires the techniques of mathematical programming and a discussion of what constitutes an optimal solution to the problem. However, a few important results can be obtained very simply.

First, we observe that any structure in the maintainable region can be attained — or approached arbitrarily closely. To see this consider the case of recruitment control. In Section 12.5 we showed that, under the operation of a constant Q the structure would converge, in the course of time, to a limiting structure satisfying

$$x(\infty) \;=\; x(\infty) \, Q = x(\infty) \, \{P + w'r\} \tag{20}$$

Hence if we wish to converge on x^* we can do so by choosing r to satisfy

$$x^* = x^* P + x^* w' r$$

i.e.

$$r = x^* (I - P)(x^* w')^{-1} \tag{21}$$

This r certainly has non-negative elements if x^* is in the maintainable region, by (15).

Control can, of course, be exercised sequentially meaning that r can be changed at each step. We might therefore expect to do rather better than sticking to the fixed strategy of (21). How might we formulate the problem of finding a better strategy? What we require is a sequence of recruitment vectors $\{r(T)\}$ such that the structure moves from $x(0)$ to x^* in an optimal fashion. 'Optimal' could be taken to mean 'as quickly as possible', 'as cheaply as possible' or 'as smoothly as possible'.

In practice we might not have unlimited time to achieve the goal as this formulation requires. We might then aim to get as near as possible to x^* in a given time. One way of implementing this idea is to get as near to x^* as possible in one step. A further step can then be taken with the same objective and so on until the time available is exhausted. The possibility of a solution along these lines depends upon being able to agree on a measure of 'distance'. A fairly general distance function which has been studied in some detail is

$$D = \sum_{i=1}^{k} W_i |x_i^* - x_i|^a , \; a > 0, \; W_i > 0 \text{ for all } i \tag{22}$$

Choice of the W_i's allows us to give some grades more weight than others and the exponent a governs the importance attached to large deviations in any grade. The problem is now to find an r which will take us from $x(0)$ to a point $x(1)$ such that the distance from $x(1)$ to x^* as measured by (22) is as small as possible. This is a problem in mathematical programming and it turns out to have a remarkably simple and appealing solution. The reader should attempt to solve the problem with $k = 2$ and $k = 3$ and $a = 1$ and $a = 2$ when it can be represented geometrically (*see* Problem (7)).

12.8. Concluding remarks

The model of a manpower system which has formed the basis of our discussion is, of course, over-simplified. Wastage rates, for example, cannot always be treated as constant within a grade. All rates are prone to change over time and in some circumstances it may be possible to project these changes. One of the great attractions of the Markov model is that it can be readily adapted to include generalisations of this kind without changing its basic structure. Hence the approach exemplified in this chapter is one which remains valid under a much wider range of conditions than the special cases we have discussed in detail.

We have drawn a distinction between the use of the model for prediction and for control. In the former case the assumptions made must reflect as accurately as possible the actual behaviour of the system in the recent past. In the case of control the assumptions fall into two groups. Those relating to the uncontrollable aspects of the system must, as for prediction, reflect what actually happens. Those relating to the control variables have a different character; they are assumptions about what it is possible for management to do and so must be based upon knowledge of how the system is organised.

12.9. Bibliography

BARTHOLOMEW, D.J. (1973). *Stochastic Models for Social Processes,* 2nd edn. Wiley; New York

12.10. Problems for further study

1. Show that the matrix Q, defined by (5), has all of its row sums equal to unity as asserted in the text.

2. Compute the grade structure of the system with parameters as given in Section 12.5 for 5 and 10 years ahead and so verify the claims made in the text.

3. Obtain the limiting grade structure for the previous problem and compare it with the projections made for 5 and 10 years ahead.

4. Find the vertices of the maintainable region for the example of Section 12.5; plot the region and compare it with the example illustrated in *Figure 12.1.*

5. Attempt to determine the maintainable region for control by promotion when $k = 3$, promotion is into the next higher grade and there is no demotion. Hence justify the remarks made at the end of Section 12.6.

6. Try to express the optimality criteria proposed in Section 12.7 for recruitment control ('as quickly as possible', 'as cheaply as possible', 'as smoothly as possible') in mathematical form.

7. Solve the 'one-step' problem of Section 12.7 using the distance function of (22) with $k = 2, 3$ and $a = 1, 2$. (Hint: consider the problem geometrically.)

8. There are obvious similarities between the stocks and flows of employees in a firm and the dynamics of human populations. For example, birth and immigration take the place of recruitment, and deaths and emigration correspond to wastage. Explore this similarity more closely and consider whether the model described in this chapter (or some modification of it) might be useful in the study of population projection and control.

13
A MATHEMATICAL MODEL FOR MOTOR INSURANCE

R.E. Beard
*Department of Trade and Industry and the Department of Mathematics,
University of Essex*

[Prerequisites: probability theory and statistics]

13.1. Introduction

For more than 200 years actuaries have been using mathematical models
as part of the scientific management of life insurance operations and of
pension funds.   In recent years considerable interest has been shown by
them in other branches of insurance and a recent symposium on
'Mathematics in Actuarial Work' (Beard, 1972) covered a representative
selection of current applications.   This chapter illustrates an insurance
problem amenable to mathematical modelling.

Insurance is basically a distribution operation within the economic
structure and its underlying principles are closely linked with social
philosophy and legal background.   It involves fairly long-term considerations and care is needed in devising models to provide for changes in
the business environment.   Indeed the value of such models is to formulate general guidelines for business policy which management will
determine on the basis of their view of external changes relative to
the assumptions made in the model.

This chapter is devoted to one aspect of the financial management
and control of motor insurance.   The discussion is limited to the case
of a limited liability company transacting UK motor business only.   The
restriction on corporate structure mainly means that operations through
Lloyds underwriters are not dealt with but as the essential principles
are similar the omission is relatively unimportant.   The restriction to
UK motor business does, however, mean the omission of a number of
important extensions arising from the transaction of other classes of
business.

The emphasis will also be on the overall operation rather than the detail arising from the variations between individual insurances, in other words the macro- rather than the micro-structure. This means there is no discussion on the very wide field of premium calculation.

13.2. Criterion for non-ruin

The most elementary model of an insurance operation is an initial fund which is augmented by the receipt of premiums and diminished by the outgo on claims. If U_0 denotes the initial fund and P_t and C_t are the premiums received and the claims incurred during an interval t then the fund at the end of the interval is

$$U_t = U_0 + P_t - C_t \tag{1}$$

If the claims outgo during the interval exceeds the initial fund and the premiums then U_t will be negative, i.e. the resources will be exhausted and the company insolvent — in risk theoretical terminology it will be ruined. The behaviour of U_t is thus a central problem and it is apparent that financial control, which is concerned *inter alia* with the continuance of the enterprise, must be concerned with the probable future behaviour of U_t and that 'static' observations of U_t, as shown for example by a balance sheet, are of limited value.

13.3. Compound Poisson process

To investigate U_t it is necessary to define the model more precisely, in particular the quantities P_t and C_t. Consider then a portfolio of N identical risks exposed to claims during a year and assume that the expected number of claims in a unit of time dt is μdt so that the total number of expected claims is $\mu N = n$, say. If it is assumed (a) that claims in two disjointed time intervals are independent, (b) that the number of claims in a time interval t_1, t_2 is dependent only on the length of the interval and (c) that multiple claims are excluded, then the claim number process will be Poisson, i.e. the probability that k claims will arise in the year is

$$P_k = \frac{e^{-n} n^k}{k!} \tag{2}$$

If the distribution function of the amount of one claim is given by $S(x)$ then the expected value of a claim is

$$\int_0^\infty x \, dS(x) = m$$

say. For a 'fair' premium the expectations are equated write, if π denotes the net risk premium

$$\pi = \mu m \tag{3}$$

and the total premiums in the year will be equal to

$$\mu \times N \times m = n \times m \tag{4}$$

If we now consider the situation when k claims have arisen we can find the distribution function of the total amount of those claims by forming the kth convolution of $S(x)$ (Beard, Pentikainen and Pesonen, 1969), i.e.

$$S^{k*}(x) = \int_0^x S^{(k-1)*}(x - z)dS(z) \tag{5}$$

By combination with the probabilities of 0, 1, ..., k ... claims we then find for the distribution function of total claims

$$F(x, n) = \sum_{k=0}^{\infty} \frac{e^{-n} n^k}{k!} S^{k*}(x) \tag{6}$$

We now write formula (1) as

$$U_1 = U_0 + nm - x$$

where x is a random variable defined by Equation (6).

The probability that $U_1 \geqslant 0$, i.e. that the company is not ruined at the end of the year is thus the probability that $x \leqslant U_0 + nm$, i.e. $F(U_0 + nm, n)$. If the desired level of non-ruin probability is fixed as $1 - \epsilon$, we have the equation

$$1 - \epsilon = \Pr \{x \leqslant U_0 + nm\} = \int_0^{U_0+nm} dF(x,n) = F(U_0 + nm,n) \tag{7}$$

13.4.   Limitations of the model

Before proceeding there are a number of points to be noted. It will be appreciated that the assumption that the order of events within the year is immaterial is unrealistic since a claim occurring on the 1st January and requiring settlement could lead to an immediate ruin situation. The solution of this problem, i.e. that the accumulated income always exceeds the claims outgo, has been a central problem in risk theoretical studies for many years and is not incorporated in this particular model. An analysis will be found, for example, in Beard, Pentikainen and Pesonen (1969), but the main results of this chapter are valid provided t is not small and ϵ is small. It is also useful to note that if in Equation (1) we are considering this 'continuous testing' model the quantity $P_t - C_t$, usually referred to as the net gain, will be a random variable with a mean of zero. For a given value of U_0

there is a positive probability that the net gain will exceed $-U_0$ so that the probability of ruin, when t increases indefinitely, is unity. To give a ruin probability less than 1 calls for a 'security loading' λ to be added to the risk premium.

It should also be noted that even if $S(x)$ is continuous at $x = 0$, there is a spike of probability at $x = 0$ (since $p_0 = e^{-n}$) which must be included in the distribution function $F(x, n)$.

It has also been assumed that μ, the claim intensity function, remains constant throughout the year. In general, this is an unrealistic assumption and 'fluctuating basic probabilities' have been introduced as a generalisation. If μ is assumed to be distributed as an incomplete γ-function then it can be shown that instead of the Poisson process the claim numbers are distributed in a negative binomial form, usually referred to as a Polya process (Haight, 1967). The main effect in the present context is to increase the amount of initial capital U_0 required.

Returning now to Equation (7) if the premiums are merely the expected claims, the company will have no resources, on the average, to meet the service on the capital U_0. Thus instead of the 'anti-ruin' margin in the continuous testing case, we must include an equivalent loading to meet the capital service charges, and the equation becomes

$$1 - \epsilon = F \{U_0 + nm (1 + \lambda), n\} \tag{8}$$

In principle, given ϵ and the distribution function $F(\cdot)$, λ can be found from Equation (8). However, marketing considerations will enter into the picture and there may be an upper limit fixed by management (there is, of course, a further line of development since n, the expected number of claims will be partly dependent on λ as the level of this will determine the competitive position of the company).

13.5. Reinsurance

If the selected values of ϵ and λ lead to an incompatible situation, it will be necessary for the company to consider some form of reinsurance. In motor insurance this usually takes the form of excess of loss reinsurance under which the reinsurer agrees to pay the amount of individual claims in excess of a given figure (the excess point) in return for a premium usually expressed as a percentage of the cedants premium income. The effect of this will be to reduce the income of the cedant and to modify the function $F(\cdot)$ by relieving him of the excess claims.

We can express this situation by the equation

$$1 - \epsilon = F^M \{U_0 + nm (1 + \lambda)(1 - r), n\} \tag{9}$$

where M denotes that the distribution of total claims $F(\cdot)$ is modified by truncation of the distribution of one claim $S(x)$ and r is the rate of reinsurance premium.

The problem now considered is to find the value of the excess point of the reinsurance treaty which solves Equation (9).   It should be remarked that other criteria can be established, such as maximising profits, or a utility function can be introduced (see, e.g. Beard, Pentikainen, Pesonen, 1969, p.160) and, of course, the result is subject to the limitations of the model in respect to the ordering of claims and the constancy of μ.

13.6.   A numerical example

In order to provide some idea of the magnitude of the various quantities involved and of the nature of the statistical complexities, a hypothetical model has been constructed for which the basic parameters are representative of practical conditions.   The various quantities are developed in the order in which they appear.   It should also be noted that the experience of each office is particular to that office so that generalisations based on collective industry figures can be very misleading.

The central function is the distribution function of one claim, $S(x)$. Suitable observational material is not only very scarce but is complicated by the effects of heterogeneity and inflation.   For UK comprehensive motor policies the cover provided consists of three main sections (a) third party personal injury, (b) third party property damage, and (c) accidental damage to the insurer's own vehicle.   These three components have significantly different shapes so that if for the purposes of statistical record all payments under a policy arising out of one accident are aggregated as one claim, the resulting distribution will be a mixture.   There are, however, more difficult aspects arising from the long time that large injury claims take to settle (with a very troublesome estimation problem) and the fact that the amount of a claim is related to the value of money at the date of settlement and not at the time of occurrence.   The distribution has a very long tail only moderately well represented by a log-normal or a Pareto type curve.   For purposes of illustration, it has been assumed that the function can be represented by the sum of a number of decreasing exponential terms and the parameters are appropriate to a constant value of money, i.e. no allowance is made for any changes in the future rate of inflation.

In practice a proportion of the claims notified to the company will result in no payment being made (for example, advice may be precautionary).   For a full analysis the proportion of nil claims should be treated as a random variable but in this example such claims have been eliminated so that the claims intensity μ, the expected number n and the function $S(x)$ relate to non-zero claims.

The numerical values adopted are

$$1 - S(x) = 0.003e^{-0.0002x} + 0.330e^{-0.005x} + 0.667e^{-0.015x}$$

with moment functions

Mean 1.2547×10^2

μ_2 1.6659×10^5, μ_3 = 2.2024×10^9, β_1 = 1.0492×10^3

σ 4.0815×10^2, μ_4 = 4.3891×10^{13}, β_2 = 1.5816×10^3

Coefficient of variation = 3.2530.

The form of $S(x)$ is representative of practical conditions and the average value of about £125.00 is appropriate to claims arising at the end of 1972: the coefficient of variation is typical.

Values of the distribution function for selected amounts are:

x	Proportion of claims not exceeding x
£0	0.00000
£100	0.64809
£500	0.96983
£1 000	0.99532
£2 000	0.99798
£5 000	0.99890
£10 000	0.99959
£15 000	0.99985
£20 000	0.99995

which will give some idea of the nature of the distribution. It should be noted that the components of the function have been determined by trial and error to provide a reasonable overall representation and are not related to the various sections of the cover.

The claim intensity μ for non-zero claims is assumed to be 0.125. A representative proportion of zero claims would be about 23% (although this figure varies substantially between companies) giving a total intensity of 0.1625. N, the number of exposures ($\equiv$ policies) is taken as 16 000 (a rather small portfolio), giving rise to 2600 expected claims, of which 2000 (= n) involve some payment. The expected amount of claims is £250 940 which is equal to the total net risk premiums.

The next stage is to deal with Equation (9). All the quantities except the excess point are given and a value has to be found which satisfies the equation. An equivalent calculation is to take ϵ as fixed and find the values of y in the equation $1 - \epsilon = F^M(y, n)$ for different values of the excess point and compare these y values with the resources $U_0 + nm (1 + \lambda)(1 - r)$. This can be done with sufficient accuracy for present purposes by using the so-called 'N-P' method (Beard, Pentikainen and Pesonen, 1969, p.43) which is derived from the Edgeworth expansion of an arbitrary distribution function in terms of the normal distribution and derivatives thereof. The moments of $F^M(x, n)$ are required and these can be derived simply from those of

$S^M(x)$ by the formulae

$$\mu_1 = n \, v_1{}^1$$

$$\mu_2 = n \, v_2{}^1$$

$$\mu_3 = n \, v_3{}^1$$

$$\mu_4 = n \, v_4{}^1 + 3 \, n^2 \, (v_2{}^1)^2$$

where the μ's are moments of $F^M(x, n)$ about the mean and the v's are moments of $S^M(x)$ about the origin.

The procedure is to first find z where $1 - \epsilon = \Phi(z)$, i.e. for $\epsilon = 0.001$, $z = 3.0902$. Then $z^1 = z + \frac{1}{6} \gamma_1 (z^2 - 1)$ is calculated where $\gamma_1 = \mu_3/\sqrt{(n\mu_2{}^3)}$ and the value of y follows from $n\mu_1 + z^1\sqrt{(n\mu_2)}$. Representative figures are:

Excess point	$n\mu_1$	z^1	$n\mu_1 + z^1\sqrt{(n\mu_2)}$	
£1 000	2.254×10^5	4.025	2.692×10^5	
£2 000	2.308×10^5	4.123	2.854×10^5	
£5 000	2.399×10^5	4.102	3.054×10^5	
£10 000	2.468×10^5	4.049	3.197×10^5	
£15 000	2.494×10^5	4.029	3.248×10^5	
£20 000	2.504×10^5	4.022	3.267×10^5	
∞	2.509×10^5	4.018	3.277×10^5	

To calculate the expected resources an initial capital (U_0) of £50 000 has been taken and a loading (λ) of 10% of the risk premiums concerned so that $nm(1 + \lambda) = £276.034$.   From this the reinsurance premiums have to be deducted and these have been taken as the expected cost of excess claims increased by a factor of $1/0.75$ to provide for reinsurers expense loadings and by a constant of £1.00 per expected claim to provide a risk loading.   There are a number of possible bases for fixing reinsurance premiums but the important feature is the market price which can be obtained and the above basis should not be regarded as other than illustrative.   One final item remains to be determined, namely the figure which should be used in this model for the service charge on the capital of £50 000.   Insurance operations provide funds the excess of which over the necessary provision of cash, financing of agency balances etc. is available for investment.   If it is assumed that investment earnings on the funds ear-marked for policyholders, i.e. premium reserves and claims provisions, are reflected in the premium calculation or otherwise, then the capital will earn some interest towards its own service.   However, investors in an insurance operation would look for a greater return than would be obtainable in pure investment so that in the financial management some regard should be had to this expectation.   For this model it has been assumed that 6% on the capital is decided upon as the appropriate figure.

The expected resources at the end of the year can now be determined as follows:

Excess point	Cap. + premiums + loading $(U_0 + P(1 + \lambda))$	Reinsurance premiums (Pr)	Resources $(U_0 + (1 + \lambda)P$ $- Pr - 0.06U_0$	F for $\epsilon = 0.001$
£1 000	£326 034	£36 000	£2.870 X 10^5	£2.692 X 10^5
£2 000	£326 034	£28 840	£2.942 X 10^5	£2.854 X 10^5
£5 000	£326 034	£16 660	£3.064 X 10^5	£3.054 X 10^5
£10 000	£326 034	£7 460	£3.156 X 10^5	£3.197 X 10^5
£15 000	£326 034	£4 000	£3.190 X 10^5	£3.248 X 10^5
£20 000	£326 034	£2 660	£3.204 X 10^5	£3.267 X 10^5
∞	£326 034	0	£3.230 X 10^5	£3.277 X 10^5

In the last column have been inserted the figures previously found for the 0.001 ruin probability level. With no reinsurance (excess point = ∞) the resources fall below the F value indicating that the ruin probability is > 0.001. With reinsurance fixed at an excess point of £1000 the resources exceed the F value showing that the reinsurance is at a higher level than needed. Equilibrium is reached at an excess point between £5000 and £10 000 which is the indicated reinsurance level on the specified conditions.

13.7. References

BEARD, R.E. (1972). 'Mathematics in Actuarial Work', *Bull. I.M.A.*, 8, No.1, 3

BEARD, R.E., PENTIKAINEN, T. and PESONEN, E. (1969). *Risk Theory.* Methuen; London

HAIGHT, F.A. (1967). *Handbook of the Poisson Distribution.* Wiley; New York. [N.B. The relevant formula 3.6 − 25 on p.46 should read

$$\mu_4 = \lambda v_4' = 3 \lambda^2 (v_2')^2]$$

13.8. Problems for further study

1. Determine the distribution function $F(x, n)$ of total claims when the claim process follows the Poisson Law and the distribution $S(x)$ of one claim is $1 - e^{-x}$.

2. Find the moments of the distribution function of total claims when the claim intensity function μ is distributed as

$$U(\mu) = \frac{1}{\Gamma(k)}\int_0^{\mu k} e^{-z} z^{k-1} dz$$

3. Calculate the approximate value of U_0 required to secure a 1 year ruin probability $\epsilon = 0.001$ when the value of k in Problem (2) = 100. Ignore reinsurance and assume $\lambda = 0.1$.

4. The moments of $F(\cdot)$ are mean = 4, μ_2 = 4, β_1 = 1, β_2 = 4.5. Compare the true value of $1 - F(x)$ for x = 4, 5, 6, 7 and 8 with the approximate values given by the $N-P$ method.

5. As a result of adverse claims experience, considered to be a random fluctuation, a company finds that its free resources (capital) have fallen below the level indicated by the desired ruin probability. Consider the effect on premium rates etc. of the three corrective measures (a) additional capital, (b) adjustment of the reinsurance excess level and (c) merger with another similar company. What considerations arise if the adverse experience is a permanent feature rather than a fluctuation?

14
A MILITARY APPLICATION OF GAME THEORY

W. Hill
AUWE, Portland, Dorset

[Prerequisites: probability theory and matrices]

14.1. Introduction

There are many situations in which people compete with one another.
In some, such as parlour games, the rules are explicitly stated. The
choices open to each player are known and the rewards are pre-deter-
mined. In many others, such as war or business, the position is much
less well defined. The possible courses of action are usually difficult
to enumerate and the aims are often multiple and complex. In all
cases, however, the aims of one person or one side run counter to
those of others. It is this element of competition which is basic to
those situations covered by the Theory of Games.

As in any theory in applied mathematics, the Theory of Games
defines a formal structure. It assumes that all the possible courses of
action of each competitor can be specified, and that for each combination of these the expected outcome of the competition can be determined numerically. The aim of each competitor is to maximise his
expected gains. All he can assume is that each opponent is also
performing a similar optimisation.

It is perhaps surprising that over a large class of problems an optimisation process satisfying these criteria can be determined in a consistent
and logical manner. That this was possible was shown by von Neumann
(1928). The resulting theory has been developed to provide a number
of far-reaching results. Although problems usually attend even successful applications, it has had some notable successes particularly in the
military field. In addition to providing an appreciation of the subtlety
which is present, even in simple competitive situations, it has thrown

light on strategic principles of attack and defence (Dresher, 1961). It has also made possible the rational design of some aspects of weapon systems and has enabled military tactics to be developed in a number of diverse fields.

In most practical problems a complete solution is not provided by the application of a single theory or a single approach. A large study often develops in several phases each requiring special techniques. The Theory of Games is just one such technique which cannot be expected to give a complete answer to large and complex problems. Nevertheless, when used properly it can make a significant contribution in many studies of conflict.

In this chapter some aspects of the theory will be developed by applying it to a problem of naval mine warfare. The aims are twofold. The first is to give the reader an appreciation of the theory and how it can be used. The second is to demonstrate how a practical problem can be solved by commencing with a simple, very artificial model, and refining it until an adequately realistic formulation is obtained.

14.2.   Two-person games

The mine warfare example chosen is one of the class of *two-person* games in which there are only two opposing sides. The most successful applications of the theory have been achieved in this category. Problems of conflict involving *n*-persons, where $n > 2$, are made considerably more difficult by the possibility of coalitions.

14.3.   Measures of effectiveness

In addition to defining all the possible courses of action (or *pure strategies*) which each side can use, it is necessary also to specify the gains each will receive for a given set of opposing pure strategies. These gains are termed *pay-offs*, and are measures of the effectiveness of pursuing the given policies. It is certainly possible for each side to obtain unrelated gains as the outcome of an encounter or 'play', or even that each side can choose a different measure. It is generally assumed, however, that each will adopt the same measure, and that what one side wins the other loses. Such games are called *zero-sum* games.

Finding the best pay-off function is often the most difficult aspect of Game Theory. There are usually many different criteria that can be used. Some may be qualitative rather than quantitative. Furthermore the relative importance of contributing factors is usually not strictly defined. Conceptually it should be possible to arrive at a single measure, or *utility* (von Neumann and Morgenstern, 1947). In the application of Game Theory a single measure must be derived, as only a single pay-off is permitted.

14.4. Mine warfare

In order to pursue the example in Naval mine warfare, it will be
necessary to describe briefly the principal elements of mines and mine
counter-measures.

The most commonly used mine is the ground influence mine. It is
laid on the sea-bed and is actuated by a change in some physical
phenomenon caused by a ship passing close-by. The usual method of
preventing damage to the ships is to employ minesweepers to explode
the mines prematurely. When sweeping influence mines, the minesweepers tow large sources which create similar changes to the physical
phenomenon as do the ships but over a much wider path. In order to
prevent the mines from being swept too easily, the mine designer uses
a device known as a ship-counter. This prevents the mine from explod-
ing until it has been actuated (by either a ship or a sweep) a pre-set
number of times. If the mine is set to explode when it receives the
nth actuation it is said to be set on ship-count n.

Although there are a number of types of mine counter-measures
operations, we shall be concerned with only one of these, namely
operations of limited duration. In these only a limited amount of
time can be spared for mine counter-measures before passing the ships
through the minefield. Both the minelayer and the mine counter
measures commander have a problem. The layer knows that, if the
ship-counter is set too low, the mines will be easily swept, and, if it
is set too high, insufficient actuations will be caused.

Thus he must aim to set the ships counters on values which offer
the best chance that the sweepers will reduce the ship counter to one
and so be ready to explode under the ships. The mine counter-
measures commander will recognise that the mine layer might well
have considered such aspects, and that if he swept a channel systematically he might play into the layer's hands. Both opponents have a
number of alternative courses of action. Instinctively, one might
expect something to be gained by each by not using obvious tactics.
This is a typical Game Theory situation.

14.5. The basic game

Initially we suppose that there is only one mine in the channel to be
swept and that only one ship will use the channel after this operation
has been completed. We shall take the pay-off to be the probability
of the ship being sunk. It is a simple and convenient measure which
summarises the effectiveness of the tactics on both sides and is also
likely to be acceptable to the military man. Using it as a pay-off
implies that the effectiveness of the operation is linearly related to it,
as utility may be arbitrary to this extent.

We define F_{ij} as the probability that a mine on ship-count i will
sink a ship passed through the channel after j passes of a sweeper

have been made along the channel. The matrix (F_{ij}) is then the pay-off matrix for the game. The optimum courses for both sides have now to be determined. The opponents are not restricted to choosing their strategies in a deterministic way. They may choose them in some random, unpredictable manner according to a selected probability distribution over their pure strategies. Thus the choice of each opponent is not necessarily a single pure strategy, but is a probability distribution over all their strategies.

Suppose that the maximum number of ship-counts that can be set on the mine is n_c and the maximum number of passes that can be carried out in the available time is n_p. Then if the minelayer chooses a probability distribution over his active strategies such that the probability of ship-count i being used is p_i, then the probability of a ship being sunk after j sweeper passes is given by

$$F_{.j} = \sum_{i=1}^{n_c} p_i F_{ij}$$

where $p_i \geqslant 0$ for all i, and

$$\sum_{i=1}^{n_c} p_i = 1$$

(Note that the dot subscript implies summing over the appropriate dimension.)

Similarly if the mine counter-measures commander determines the number of passes at random from a probability distribution giving the probability of making j passes q_j, then the probability of sinking, for a mine on ship count i is

$$F_{i.} = \sum_{j=1}^{n_p} F_{ij} q_j$$

where $q_j \geqslant 0$ for all j, and

$$\sum_{j=1}^{n_p} q_j = 1$$

The probability distributions $P = (p_1, p_2, ..., p_{n_c})$, $Q = (q_1, q_2,..., q_{n_p})$ are called *mixed strategies.* When each side uses mixed strategies the probability of sinking is

$$F_{..} = \sum_{\text{all } i,j} \sum p_i F_{ij} q_j$$

Evidently this formulation includes those cases where only one pure strategy is used.

The aim of the minelayer is to maximise $F_{..}$, and that of the mine counter-measures commander is to minimise it. Suppose the minelayer chooses a mixed strategy $P^* = (p_i{}^*)$. Then he will be sure of achieving a probability of sinking of at least

$$\min_{Q} \sum_i \sum_j p_i^* F_{ij} q_j$$

where the mine counter-measures commander is assumed to have determined the distribution Q that minimises the summation. Now the minelayer can choose the distribution P^* to be that one, say $\hat{P} = (\hat{p}_i)$, that maximises its value. Thus he can ensure that the probability of sinking is at least

$$\max_{P} \min_{Q} \sum_i \sum_j p_i F_{ij} q_j$$

i.e. that

$$F_{o.} = \sum_i \sum_j \hat{p}_i F_{ij} q_j \geqslant \max_{P} \min_{Q} \sum_i \sum_j p_i F_{ij} q_j$$

(Note that the suffix o indicates the use of the maximising strategy P.)

Similarly it can be shown that the mine counter-measures commander can use a strategy $\hat{Q} = (\hat{q}_j)$ and ensure that the probability of sinking is never more than

$$\min_{Q} \max_{P} \sum_i \sum_j p_i F_{ij} q_j$$

i.e. that

$$F_{.o} = \sum_i \sum_j p_i F_{ij} \hat{q}_j \leqslant \min_{Q} \max_{P} \sum_i \sum_j p_i F_{ij} q_j$$

It can be easily shown that

$$\max_{P} \min_{Q} \sum_i \sum_j p_i F_{ij} q_j \leqslant \min_{Q} \max_{P} \sum_i \sum_j p_i F_{ij} q_j$$

The fundamental theorem of matrix games shows, however, that these two expressions are always equal (von Neumann, 1928; Dresher, 1961; von Neumann and Morgernstern, 1947; McKinsey, 1952). Therefore

$$F_{.o} \leqslant F \leqslant F_{o.} \tag{1}$$

$F = F_{oo} = \sum_i \sum_j \hat{p}_i F_{ij} \hat{q}_j$ is called the 'value' of the game and the mixed strategies $\hat{P}$ and $\hat{Q}$ are called optimal strategies.

Thus we have shown how in a simple one mine, one ship game, the optimum policies for the minelayer and the minesweeper can be determined. Each can ensure that the result is never worse than F (from his point of view), and if his opponent uses a poor strategy he can do better. It must be realised however that these results are expected values which apply *before* the actual strategies to be played have been obtained from the chance device. After they have been determined the result will be given by the appropriate pay-off in the game matrix. In general this will be different from the game value.

An exception occurs when the optimal strategies are pure strategies, that is when $\hat{p}_i$ and $\hat{q}_j$ are unity for one value of i, j.   This is called a 'saddle-point' solution.   When it occurs each player behaves in a deterministic manner, knowing that nothing is gained from concealing his intentions.   Saddle-point solutions are characterised by the fact for them

$$\max_i \ \min_j \ F_{ij} \ = \ \min_j \ \max_i \ F_{ij}$$

When solutions of games are required the first step is to check whether a saddle-point solution exists.   If it does not then more elaborate methods are required.   These include graphical methods for $2 \times n$ games, matrix methods, iterative methods and linear programming methods.   They will be found in standard works on Game Theory, Linear Programming and Operational Research (McKinsey, 1952; Dantzig, 1963; Hillier and Liberman, 1966).

14.6.   First extension: many mines

There are many artificialities in our simple mine counter-measures game, which could severely limit the practical value of the results obtained from it.   The most restrictive assumption is that there is that there is exactly one mine in the channel.   Mines are generally laid in large numbers over an area of sea, and their precise positions are unknown.   Thus any channel through the minefield will contain an unknown number of mines.

It is better therefore to assume that the mines have been laid randomly, and to estimate the density of mining.   Suppose it is such that the expected number of mines in a channel of total length Y is N.   Then the probability of a mine being in an element of length Δy is $N\Delta y/Y$.   The probability that a ship, on reaching the element, passes through safely is

$$1 - \frac{N \ F_{ij} \ \Delta y}{Y}$$

where the mines have been set on ship-count i and j sweeper passes have been made.

If $P(y)$ is the probability that the ship is undamaged when it reached y, we have that

$$P(y + \Delta y) \ = \ P(y)\left(1 - \frac{N \ F_{ij} \ \Delta y}{Y}\right)$$

Thus

$$\frac{P(y + \Delta y) - P(y)}{\Delta y} = - \frac{NF_{ij}}{Y} P(y)$$

and in the limit, as $\Delta y \to 0$,

$$\frac{dP(y)}{dy} = - \frac{NF_{ij}}{Y} P(y)$$

Since NF_{ij}/Y is a constant, this is a simple first-order ordinary differential equation. Integrating

$$P(y) = P(0) \exp\left(- \frac{NF_{ij}}{Y} y\right)$$

Now $P(0) = 1$, so the probability that the ship survives is

$$P_{ij} = \exp(-NF_{ij})$$

and the probability that it will be sunk is

$$R_{ij} = 1 - \exp(-NF_{ij}) \tag{2}$$

where R is used to denote the 'Risk'.

It will be seen that when there is uncertainty about the number of mines in the channel, the probability of sinking is not in general a linear function of its value in the single-mine case. That is, the risk, R_{ij}, is not in general a linear function of F_{ij}.

Thus, in general, the solutions to the simple game would not be the solution to this extended one. If, however, the probability of the ship being sunk is small ($NF_{ij} \ll 1$, for all i, j), as will be the mine counter-measures commanders intention, then Equation (2) can be replaced by the approximation

$$R_{ij} = NF_{ij} \tag{3}$$

R_{ij} and F_{ij} are now linearly related, and the solutions to the simple one-mine game will be applicable. By using the optimal strategy of this game the minelayer will be able to ensure that he achieves a risk of $R = NF$ or more. Similarly the mine counter-measures commander can ensure that the risk is R or less. As in the simple game, however, these results apply only before the actual pure strategies are determined by either side, unless the game has a saddle-point solution. After the strategies are determined the actual risk will be that given by the appropriate pay-off.

14.7.   Second extension: mixtures of strategies

It is readily apparent that laying all his mines on a single ship-count is not a wise policy for the layer.   Although he can obtain a comparatively high risk if the mine counter-measures commander unluckily uses a certain strategy, the random choice mechanism advocated by Game Theory will ensure that this is unlikely to happen.   The layer therefore will spread his ship-counts so as to give a greater assurance of achieving some success over a range of his opponents strategies.

Suppose he places a fraction p_i of his mines on ship count i ($i = 1, ..., n_c$).   Then, following the same reasoning as in the preceding equations for each ship count in turn, Equation (3) becomes

$$R_{.j} := N \sum_i p_i F_{ij} = \sum_i p_i R_{ij}$$

As $p_i \geqslant 0$, for all i, and

$$\sum_i p_i = 1$$

we see that the resulting risk, R_{ij}, is a convex linear combination of the risks for the individual ship-counts.

Similar reasoning would persuade the mine counter-measures commander that he should not adopt a single sweeping policy for the whole channel (unless practical considerations force him to do so).   Instead, he should spread his effort over the length of the channel non-uniformly. If he divides the channel into segments so that in a fraction q_j of the channel he sweeps to j passes ($j = 1, 2, ..., n_p$), then in segment $q_j Y$ the risk to the ship from mines on ship-count i will be $\{1 - \exp(-NF_{ij} q_j)\}$.   Thus the probability of being sunk for the whole channel will be

$$1 - \prod_j \exp(-NF_{ij} q_j) = 1 - \exp(-N \sum_j F_{ij} q_j)$$

and so in this case Equation (3) becomes

$$R_{i.} = N \sum_j F_{ij} q_j = \sum_j R_{ij} q_j$$

It can be seen that, as for the minelayer, a minesweeping strategy composed of a number of pure strategies carried out in different lengths of the channel gives a risk which is a convex linear combination of the risks for each of the pure strategies.

We are now considering a reasonably realistic situation, with the minelayer laying a mix of mines in the minefield and the minesweepers varying the density of sweeping along the channel length.   It is evident that the number of strategies now open to the minelayer is a very large number (for a reasonable number of mines laid in the field) and

is infinite for the minesweepers. Thus the resulting game is an infinite
one. Since the pay-offs for all these strategies are convex linear combinations of the pay-offs in the simple finite game, the solution of the
finite game also gives the solutions to the infinite one. The values of
the games are the same. The mixed strategy solutions of the simple
game may be interpreted in two ways. As a mixed strategy, the solution results in the same policy as in the simple game, with a single
pure strategy being adopted. If, however, the probability distributions
for the optimal strategies are used to determine the mix of mines or
sweeper passes, then a saddle-point solution results. The 'mixtures of
strategies' constitute pure strategies in the infinite game. Although
both interpretations lead to the same expected risk, the latter has the
additional property that the variability of the result obtained over a
number of hypothetical engagements is a minimum when it is used.
Thus the use of mixtures of strategies, which one would intuitively
consider to be worthwhile, has indeed an additional beneficial property.

EXAMPLES

The benefit obtained from using mixtures of strategies in an operation
can be seen by considering two simple examples. In both the minelayer can lay his mines on ship-counts 1, 2, 3, and the minesweepers
can carry out 0 or 1 pass. Then for the simple game we have the
following game (*Table 14.1*). It is evident that the matrix value of
the game is 0.1, and that the optimal strategies each with probabilities
of 0.5. In any one operation, however, the result will be either 0 or
0.2.

Table 14.1 Simple strategies, mixed strategy solution

Ship count	Passes		Proportion p_i
	0	1	
1	2	0	0.5
2	0	2	0.5
3	0	0	0
Proportion q_j	0.5	0.5	Value 0.1

2 mines in field; 0, 1 sweeper passes; 1, 2, 3 ship-counts

In *Table 14.2* we show the infinite game for the same choice of
mines and sweeper passes. It can be seen that the value of the game
is still the same, but that in this case a saddle-point solution exists,
the minelayer lays half his mines on ship-count 1, half on ship count
2, and the sweepers half the channel to 1 pass. In this case the mine
counter-measures commander can ensure that the risk to the ship is the
game value or less in every operation.

Table 14.2 Mixtures of strategies, saddle-point solution

		Proportion of channel					Proportion
Passes		$1-\Delta$	0.5	Δ	0		
		Δ	0.5	$1-\Delta$	1		
Passes	1						
	0						
Ship-count (1,1)	0.2	$0.2(1-\Delta)$	0.1	0.2Δ	0		0
(1,2)	0.1	0.1	$\boxed{0.1}$	0.1	0.1		1
(1,3)	0.1	$0.1(1-\Delta)$	0.05	0.1Δ	0		0
(2,2)	0	0.2Δ	0.1	$0.2(1-\Delta)$	0.2		0
(2,3)	0	0.1Δ	0.05	$0.1(1-\Delta)$	0.1		0
(3,3)	0	0	0	0	0		0
Proportion	0	0	1	0	0		Value 0.1

14.8. Third extension: many ships

The investigation must now be extended to consider the passage of a number of ships along the channel. Initially we shall consider the case where the expected number of ships lost in passing through the channel is the appropriate criterion for deciding the effectiveness of the operation. When a single ship is passed through the channel the only mines which can be effective are those whose ship-counter is unity after the sweeping has taken place. When more than one ship traverses the channel this is no longer strictly true. In practical circumstances, however, the chance of more than one actuation being caused by the ships is very small and will be neglected. Thus we have that the probability of a mine in the channel on ship-count i sinking one of M ships after j sweeper passes have been carried out is MF_{ij}.

To determine rigorously the number of ships sunk in passing through the channel would require the use of the theory of stochastic processes. It is possible, however, to obtain a simpler derivation which is sufficiently accurate when the losses are small. It is assumed that if M ships enter the channel, the probability of one being sunk in any element, Δy, of the channel is independent of any earlier casualties and is

$$\frac{MNF_{ij}\ \Delta y}{Y} \qquad (4)$$

Thus the distribution of the losses will be Poisson, and the expected number of ships sunk will be

$$\mu_{ij} \ = \ MNF_{ij} \ = \ MR_{ij} \qquad (5)$$

It is left to the reader to show that for both the minelayer and the minesweepers mixtures of strategies give expected losses which are convex linear combinations of the losses for the pure strategies of the finite game. The solution is given by Hill and Wallis (1970). (It should be noted that in it the rows and columns of the pay-off matrix have been transposed to conform to mine-warfare usage.)

Thus, with an expected loss criterion the solution of the infinite game obtained when each side uses a mixture of strategies will follow those for a single ship. The optimal strategies are given by those for a simple game (F_{ij}) and may be interpreted either as mixed strategies as in the finite game or as a mixture of strategies resulting in a saddle-point solution. The value of the game will be MNF. As in the single-ship case it can be shown that the interpretation as a mixture of strategies gives the smallest variability in actual outcome over repeated plays.

14.9. Final extension: low loss criterion

Uncertainty about the outcome of military operations is almost inevitable. It is often not of over-riding significance in a long campaign where many similar engagements take place. High losses in some engagements are offset by low losses in others. Thus in very many cases, although variability is present, it is the expected result in an engagement which is the appropriate measure of its success. Naturally there will be a desire to reduce the variability as much as possible. A method for doing this by using mixtures of strategies has been shown in the preceding paragraphs.

In some instances, however, a particular operation is of crucial importance and the outcome of a whole campaign might well depend on it. Under these circumstances a mean or expected outcome is not a particularly satisfactory criterion, as the variability about this value might be high. For example in the passing of a number of ships through a cleared channel the losses follow a Poisson distribution. It is a property of this distribution that the mean and the variance are of equal magnitude and thus significant departures from the mean value can be obtained. Thus there could be a considerable risk of the operation failing even though the mean value would have achieved success. In these cases the possible variability must be taken into account when determining what measure of effectiveness should be used.

This situation can arise in the mine warfare game we have been studying. Many mine counter-measures operations will have failed if more than a certain percentage of the ships are sunk in passing through the minefield. Thus in these operations the expected number of ships sunk is not an adequate criterion of effectiveness.

It has been shown that providing the losses are small, the probability distribution of the losses is approximately Poisson. Then in the case of the simple game with matrix (F_{ij}), the probability of obtaining m or more losses from the M ships passing along the channel is given approximately by

$$P_{ij}(m) = 1 - \exp(-MNF_{ij}) \left\{ 1 + MNF_{ij} + \frac{(MNF_{ij})^2}{2!} + \dots + \frac{(MNF_{ij})^{m-1}}{(m-1)!} \right\}$$

It will be seen from this expression that $P_{ij}(m)$ is not a convex linear combination of the F_{ij}, and so the simple game will not provide the solution to the finite game with matrix $(P_{ij}(m))$. To obtain a solution the correct game must be played.

Let us now consider again the infinite game where each side may employ mixtures of its pure strategies in any one operation. Let $P_{..}(m)$ denote the probability of m or more ships being sunk when the mixed strategies for the simple game are used as mixtures of strategies for the infinite game. Then $P_{..}(m)$ is given by

$$P_{..}(m) = 1 - \exp(-MNF_{..})$$
$$\left\{ 1 + MNF_{..} + \frac{(MNF_{..})^2}{2!} + ... + \frac{(MNF_{..})^{m-1}}{(m-1)!} \right\}$$

It can be seen that $P_{..}(m)$ is not a convex linear combination of the constituent values of $P_{ij}(m)$. Thus the finite game with pay-offs $P_{ij}(m)$ will not in general result in a solution to the infinite game.
If we consider the function

$$\theta = 1 - \exp(-\phi)\left\{ 1 + \phi + \frac{\phi^2}{2!} + ... + \frac{\phi^t}{t!} \right\}$$

we have

$$\frac{d\theta}{d\phi} = \exp(-\phi)\left\{ 1 + \phi + \frac{\phi^2}{2!} + ... + \frac{\phi^t}{t!} \right\}$$

$$- \exp(-\phi)\left\{ 1 + \phi + \frac{\phi^2}{2!} + ... + \frac{\phi^t}{(t-1)!} \right\}$$

$$= \exp(-\phi)\frac{\phi^t}{t} > 0, \text{ for } \phi > 0$$

Thus when $\phi > 0$, θ is a strictly increasing function of ϕ. It follows that P_{ij} is a strictly increasing function of F_{ij}. The more exact analysis based on the theory of stochastic processes, which has been mentioned above, may be used to confirm this result.
Now we have from the solution to the game with matrix (F_{ij}) that

$$\sum_i \sum_j \hat{p}_i F_{ij} q_j \geqslant \sum_i \sum_j \hat{p}_i F_{ij} \hat{q}_j \geqslant \sum_i \sum_j p_i F_{ij} \hat{q}_j \tag{6}$$

where again $\hat{p}_i$, $\hat{q}_j$ are the optimal strategies. If we denote by $P_{o.}$, $P_{..}$, $P_{.o}$ the values of $P_{..}$ when the strategy mixtures in the infinite game are $(\hat{p}_i, q_j)$, $(\hat{p}_i, \hat{q}_j)$, $(p_i, \hat{q}_j)$ respectively, then from Equation (6), and because $P_{..}$ is a strictly increasing function of $F_{..}$ we have

$$P_{o.}(m) \geqslant P(m) \geqslant P_{.o}(m)$$

It can be shown that this is sufficient to establish that $F(m)$ is a saddle-point solution of the infinite game. Thus again the solution for the simple game (F_{ij}) has provided the solution for the practical situation.

14.10.   Summary

It is worthwhile to summarise the lessons that can be learned from the example of Game Theory used in this chapter. Although it has often been said that Game Theory has not been applied to many practical situations, we have seen that in at least one type of military engagement the theory can be used to good effect. It has also been shown that there is often more than one way in which the results of a game-theory analysis can be interpreted, and that some of these interpretations are more useful than others. In particular there is some merit in achieving saddle-point solutions whenever possible.

By taking the study of the mine and mine counter-measures operation through a number of stages of increasing complexity, it has been demonstrated how the original simple concepts can be successively refined until a reasonable working model is obtained. Furthermore, the simpler studies in such cases do not represent nugatory effort but aid understanding and sometimes provide the actual solutions for the more complex cases.

14.11.   References

DANTZIG, G.B. (1963). *Linear Programming and Extensions*, Princeton University Press; New Jersey

DRESHER, M. (1961). *Games of Strategy: Theory and Applications.* Prentice-Hall Applied Mathematics Series, Prentice Hall; Englewood Cliffs, New Jersey

HILL, W. and WALLIS, P.R. (1970). 'The Application of the Theory of Games to Mine Countermeasures Tactics', *J. Inst. Math. Applic.*, **6**, 27

HILLIER, F.S. and LIEBERMAN, G.J. (1966). *Introduction to Operations Research.* Holder-Day Inc.; California

McKINSEY, J.C.C. (1952). *Introduction to the Theory of Games.* McGraw-Hill; New York

VON NEUMANN, J. (1928). 'Zur Theorie der Gesellschaftsspiele', *Math. Ann.*, **100**, 295

VON NEUMANN, J. and MORGERNSTERN, O. (1947). *Theory of Games and Economic Behaviour*, Princeton University Press; New Jersey

14.12.   Problems for further study

1. In the game of two-finger Morra the two players simultaneously show one or two fingers, and at the same time call their guess of the number of fingers shown by their opponent. If a player guesses correctly he receives from his opponent a number of points equal to the total number of fingers shown. Thus if a player shows 1 finger and calls 2, and is correct, he will win three points from his opponent. Incorrect guesses by both players result in zero score. It can easily be shown that the game matrix is as follows, where (1,2) denotes the strategy of showing one finger and calling 2.

		B			
		(1,1)	*(1,2)*	*(2,1)*	*(2,2)*
	(1,1)	0	2	−3	0
A	*(1,2)*	−2	0	0	3
	(2,1)	3	0	0	4
	(2,2)	0	−3	4	0

A is the maximising player. As the game is symmetric it is evident that its value must be zero. Show that an optimal strategy for each player is (0, 0.6, 0.4, 0).

2. With the notation in the chapter, show that $R_{.o} \leqslant R$

3. If the variance of risk in an ensemble is given by

$$\text{var}_{o.}(R) = \sum_i \hat{p}_i \, (R_{i.} - R_{o.})^2$$

show that

$$\text{var}_{o.}(R) = \sum_i \hat{p}_i \, R_{i.}^2 - R_{o.}^2$$

4. Explain the nature of the approximations made in the derivation of the Poisson distribution in Section 14.9.

5. Assuming that there is a negligible probability that the passage of M ships will actuate any mine more than once, and that the probability of a ship being sunk in an element, Δy, of channel length Y is $u\Delta y/Y$, show that the probability distribution of ship losses is given by

$$P(k) = \binom{M}{k} \exp{(-Mu)} \, \{\exp{(u)} - 1\}^k, \quad k \leqslant M$$

where $P(k)$ is the probability of k ships being sunk. (This example requires the application of the theory of stochastic processes.)

6. The only counter-measure against a new enemy weapon is ineffective against other, less effective, weapons. Should the counter-measure be developed, and if so is it likely to be used?

7. Colonel Blotto has been ordered to occupy two positions behind the enemy lines. He has four regiments under his command and the enemy commander has three. Both sides have to decide how many regiments they will assign to each post. The pay-offs to Colonel Blotto are determined in the following manner. He wins a post if he has assigned more regiments to it than the enemy has. The measure of his success is one more than the number of enemy regiments at the post. If he has assigned fewer regiments to the post than the enemy, then he loses the post. The measure of his loss is one more than the number of regiments he sent to attack it. If equal numbers of regiments are assigned by both sides the measure is zero. The total pay-off is the sum of the measures at the two posts.

Construct the pay-off matrix and solve the game.

Suppose one of the posts is more important than the other. How can this fact be included in the study?

15
NETWORK FLOW MODELS

B.A. Carré
*Centre d'Information, Université Paul Sabatier, Toulouse, France
and Department of Electronics, University of Southampton*

[Prerequisites: linear algebra and elementary analysis]

15.1. Introduction

The study of network flow models began in the 1940's, in connection
with transportation problems, i.e. problems of transporting a commodity
from certain points of supply to other points of demand, in such a
way as to minimise shipping cost.   However, it was soon appreciated
that the types of mathematical models developed in this context could
also be used to formulate and solve other network problems, concerning
for instance the flow of information in communication networks, and
the flow of traffic in road networks.   Furthermore, it has been found
that many practical problems of a combinatorial nature, which are not
associated with networks in any physical sense, can nevertheless be formulated and solved very elegantly by using network models.
  Here we shall first present the basic principles of Network Flow
Theory, and develop a method of solving an important class of flow
problems.   By means of examples, it will then be shown how network
models can be constructed and used to solve a variety of practical
problems.   An outline of some extensions of the basic theory will
then be given.
  There are often several conceptually quite different ways of constructing a mathematical model of the same physical problem, which may
suggest quite different methods of solution.   Indeed, all the problems
which will be presented here can be formulated and solved by several
radically different methods, as will be demonstrated in the final section.

15.2. Networks

A *network* (or *graph*) $G = (N, A)$ consists of

1. A set of *nodes* $N = \{n_1, n_2, ..., n_p\}$.
2. A set A of ordered pairs (n_i, n_j) of elements of N; the members of A are called *arcs*.

A network can be depicted by a diagram in which nodes are represented by points in the plane, and each arc (n_i, n_j) is indicated by an arrow drawn from the point representing n_i to the point representing n_j. For example, *Figure 15.1* depicts a network having the four nodes n_1-n_4 and the six arcs (n_1, n_2), (n_1, n_4), (n_2, n_4), (n_3, n_2), (n_4, n_1) and (n_4, n_3).

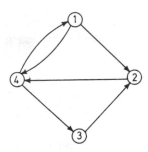

Figure 15.1

Many physical structures can conveniently be considered as networks of this kind — one example being a city street system, where we regard each 'two-way' street as a pair of contiguous one-way streets, and represent it by a pair of arcs having the same endpoints but opposite orientations.

On a network $G = (N, A)$, an arc (n_i, n_j) is said to be *directed* from n_i to n_j, and we call n_i and n_j the *initial* and *terminal* nodes of (n_i, n_j), respectively. A sequence of arcs such that the terminal node of each arc coincides with the initial node of the next is called a *path*. Thus on the graph of *Figure 15.1*,

$$(n_1, n_2), (n_2, n_4), (n_4, n_3)$$

is a path from n_1 to n_3. A *chain* is a sequence of arcs such that each intermediate arc is joined to the preceding arc at one of its extremities, and to the following arc at its other extremity. For instance on *Figure 15.1*,

$$(n_1, n_4), (n_2, n_4), (n_3, n_2)$$

is a chain from n_1 to n_3. Thus in following a chain, it is possible to traverse an arc in the direction opposite to its orientation: the arcs which are traversed in the direction of their orientation are called *forward* arcs of the chain, the others being *reverse* arcs.

15.3.   Flows in networks

A problem which frequently arises in practice is that of maximising the flow of some commodity between two specified nodes of a network, when the amount which can flow along each arc is limited. For instance, in a network of city streets, one might want a maximal traffic flow from one point to another. This problem, called the *maximal flow problem,* can be formulated in mathematical terms as follows.

Let us suppose that each arc (n_i, n_j) of a network $G = (N, A)$ has associated with it a non-negative real number $c(n_i, n_j)$ the *capacity* of (n_i, n_j). Also, let us distinguish two nodes n_s and n_t on G, which we call its *source* and *sink*, respectively. Then a *network flow* of value v from n_s to n_t is a function ϕ from A to the set of non-negative real numbers which satisfies the conditions

$$\sum_{(n_i, n_j) \in A_i^+} \phi(n_i, n_j) - \sum_{(n_j, n_i) \in A_i^-} \phi(n_j, n_i) = \begin{cases} v, & \text{if } n_i = n_s \\ 0, & \text{if } n_i \neq n_s, n_t \\ -v, & \text{if } n_i = n_t \end{cases} \tag{1}$$

$$\phi(n_i, n_j) \leqslant c(n_i, n_j), \quad \text{for all } (n_i, n_j) \in A \tag{2}$$

where A_i^+ is the set of arcs with initial node n_i, and A_i^- is the set of arcs with terminal node n_i. We call $\phi(n_i, n_j)$ the *arc flow* in (n_i, n_j). Equation (1) states that the net flow out of the source is v, that the net flow out of each intermediate node is zero, and that the net flow into the sink is v; Equation (2) states that in each arc, the flow cannot exceed the capacity. A *maximal flow* from n_s to n_t is a flow ϕ, satisfying (1) and (2), for which v is maximal.

An example of a network flow is given in *Figure 15.2,* in which the first number associated with each arc is its capacity, the second being the arc flow; bold lines indicate arcs which are *saturated,* i.e. arcs in which the flow is equal to the capacity. At first sight this network flow might appear to be maximal, but a flow of greater value can be obtained (*see Figure 15.4*).

15.4.   A method of finding maximal flows

It will be convenient to use the following notation: if P and Q are two subsets of the node-set N of a network $G = (N, A)$, then (P, Q)

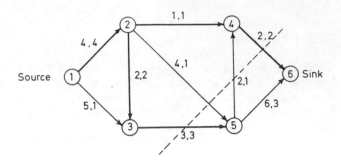

Figure 15.2   A network flow of value 5

denotes the set of all arcs with initial nodes in P and terminal nodes in Q; also, we let

$$\phi(P, Q) = \sum_{\substack{n_i \in P \\ n_j \in Q}} \phi(n_i, n_j) \tag{3}$$

and

$$c(P, Q) = \sum_{\substack{n_i \in P \\ n_j \in Q}} c(n_i, n_j) \tag{4}$$

Finally, we denote by $\bar{P}$ the complement of P relative to N (that is, the set of all elements of N which do not belong to P).

To solve the maximal flow problem we must introduce one more concept, that of a 'cut': A *cut in G separating* n_s *from* n_t is a set of arcs $(P, \bar{P})$ where $n_s \in P$ and $n_t \in \bar{P}$. (It is evident that, if one removes the arcs of a cut from G, one destroys all paths from the source to the sink.) As an illustration, the broken line on *Figure 15.2* indicates a cut separating n_1 from n_6: here $P = \{n_1, n_2, n_3, n_4\}$, $\bar{P} = \{n_5, n_6\}$, and

$$(P, \bar{P}) = \{(n_2, n_5), (n_3, n_5), (n_4, n_6)\}$$

(Notice that the arc (n_5, n_4) belongs to $(\bar{P}, P)$, but not to $(P, \bar{P})$.) We describe $c(P, \bar{P})$ as the *capacity of the cut* $(P, \bar{P})$. Thus the cut indicated on *Figure 15.2* is of capacity 9.

Now let $(P, \bar{P})$ be any cut separating the source from the sink on a network G. Then summing the Equations (1) over all $n_i \in P$, and noting cancellations, we obtain

$$v = \phi(P, \bar{P}) - \phi(\bar{P}, P) \qquad (5)$$

Since all arc flows are non-negative, $\phi(\bar{P}, P) \geqslant 0$; also $\phi(P, \bar{P}) \leqslant c(P, \bar{P})$ by virtue of (2). Hence (1) gives

$$v \leqslant c(P, \bar{P}) \qquad (6)$$

In words, (6) states that for an arbitrary flow and an arbitrary cut, the value of the flow is less than or equal to the capacity of the cut. This result is hardly surprising, but we have drawn attention to it because it has the following important implication: if we find some flow ϕ and some cut $(P, \bar{P})$ such that the value of ϕ is equal to the capacity of $(P, \bar{P})$, then we can be sure that ϕ is maximal (and that $(P, \bar{P})$ is a cut of minimal capacity). We will now demonstrate that there always exists a flow ϕ and a cut $(P, \bar{P})$ for which equality holds in (6), and give an algorithm for constructing such a flow.

If ϕ is a flow on a network $G = (N, A)$, then the *associated incremental network* $G'(\phi)$ is the network having the same nodes as G and arcs determined as follows. For each arc (n_i, n_j) of G, $G'(\phi)$ contains

1. A *normal* arc (n_i, n_j), if on G

$$\phi(n_i, n_j) < c(n_i, n_j)$$

and

$$\phi(n_j, n_i) = 0 \quad (\text{or } (n_j, n_i) \notin A)$$

2. An *inverted* arc (n_j, n_i), if on G

$$\phi(n_i, n_j) > 0$$

On $G'(\phi)$, each normal arc (n_i, n_j) has a capacity

$$c'(n_i, n_j) = c(n_i, n_j) - \phi(n_i, n_j)$$

and each inverted arc (n_j, n_i) has a capacity

$$c'(n_j, n_i) = \phi(n_i, n_j)$$

Note that each arc of $G'(\phi)$ is associated with a particular arc of G, having the same endpoints; the orientations of these two arcs are the same if the arc of $G'(\phi)$ is normal, and opposite if the arc of $G'(\phi)$ is inverted.

As an example, *Figure 15.3* shows the incremental network associated with the network flow of *Figure 15.2*. On *Figure 15.3*, inverted arcs are indicated by broken lines; the numbers on the arcs are their capacities.

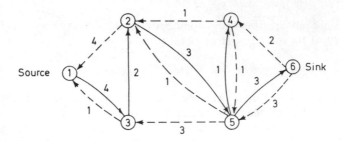

Figure 15.3   Incremental network

Now let us suppose that on $G'(\phi)$ there exists some path μ from n_s to n_t, and let ϵ be the minimum of $c'(n_i, n_j)$ taken over all arcs (n_i, n_j) of this path. Then μ determines a chain γ from n_s to n_t on G (each normal arc of μ determining a forward arc of γ, each inverted arc of μ determining a reverse arc of γ).

Furthermore, on G, we can *increase* the flow in each *forward* arc of γ by ϵ, and *reduce* the flow in each *reverse* arc of γ by ϵ, without violating the arc flow constraints (2), and without changing the net output of any node other than the source and sink; for these two nodes, the net output and input respectively are increased by ϵ units. Hence these modifications give a new network flow, whose value is ϵ units greater than the previous one.

As an illustration, the incremental network of *Figure 15.3* contains a path

$$\mu = (n_1, n_3), (n_3, n_2), (n_2, n_5), (n_5, n_6)$$

for which $\epsilon = 2$. The modification of the flows in the arcs of the corresponding chain, on the graph of *Figure 15.2*, results in the flow of *Figure 15.4*.

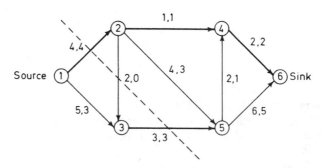

Figure 15.4   A network flow of value 7

Let us now suppose that $G'(\phi)$ does *not* contain any paths from n_s to n_t. Let P be the subset of N consisting of n_s and all nodes which are *accessible* from n_s, i.e. which lie on paths from n_s. Then since $n_s \in P$ and $n_t \in \bar{P}$, $(P, \bar{P})$ is a cut of G separating n_s from n_t. Now in every arc (n_i, n_j) of $(\bar{P}, P)$ the flow must be null, for otherwise $G'(\phi)$ would contain an inverted arc (n_j, n_i), rendering the node $n_j \in \bar{P}$ accessible from n_s. Accordingly, every arc (n_i, n_j) of $(P, \bar{P})$ must be saturated, for otherwise $G'(\phi)$ would contain a normal arc (n_i, n_j) rendering $n_j \in \bar{P}$ accessible from n_s. Hence $\phi(\bar{P}, P) = 0$ and $\phi(P, \bar{P}) = c(P, \bar{P})$, and (4) gives

$$v \;=\; \phi(P, \bar{P}) - \phi(\bar{P}, P) \;=\; c(P, \bar{P}) \tag{7}$$

It follows from (6) and (7) that ϕ is a maximal flow, and that $(P, \bar{P})$ is a cut of minimal capacity.

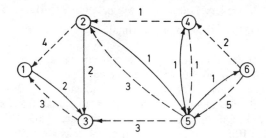

Figure 15.5   Incremental network

As an illustration, *Figure 15.5* shows the incremental network associated with the network flow of *Figure 15.4*. On *Figure 15.5* there are no paths from the source to the sink, and we have $P = \{n_1, n_3\}$, $\bar{P} = \{n_2, n_4, n_5, n_6\}$. The broken line on *Figure 15.4* indicates the corresponding cut.

$$(P, P) \;=\; \{(n_1, n_2), (n_3, n_5)\} \tag{8}$$

which has a capacity of 7, equal to the value of the flow.
We have effectively proved that

1. *A flow ϕ on a network G is of maximal value if and only if the associated incremental network $G'(\phi)$ does not contain any paths from the source to the sink.*
2. *The Max-Flow Min-Cut Theorem (Ford and Fulkerson, 1962): For any network, the maximal value of a flow from source to sink is equal to the minimal capacity of a cut separating the source from the sink.*

If we now impose a mild restriction on the arc capacities — specifically, if we assume that all arc capacities are integers — then the first of these results leads to the following simple algorithm for constructing a maximal flow:

Step 1:   Choose arbitrarily some integer-valued network flow from n_s to n_t. (The *null flow,* in which all arc flows are zero, is a possible choice.)

Step 2:   Let ϕ be the present network flow. Construct the associated incremental network $G'(\phi)$, and search for a path from n_s to n_t on this network. If such a path exists proceed to *Step 3,* otherwise go to *END.*

Step 3:   Let μ be a path from n_s to n_t on $G'(\phi)$, let ϵ be the minimal arc capacity on μ, and let γ be the chain on G determined by μ. Increase (decrease) the flow in each forward (reverse) arc of γ by ϵ, then return to *Step 2.*

END   The flow ϕ is maximal.

Clearly, if the arc capacities are integers, and the computation is initiated with an integral flow, each successive flow is integral. And since each application of *Step 3* increases the flow value by at least one unit, the algorithm will give a maximal flow after a finite number of steps. On the other hand, if the arc capacities are not integers the algorithm may not be finite, or may terminate with a non-maximal flow (Ford and Fulkerson, 1962; Hu, 1969). However, the restriction to integral capacities has no practical importance, since rational capacities can always be reduced to integral capacities by clearing fractions.

15.5. Minimal-cost flows

In general there are numerous flows from source to sink having the same flow value. (For example, there are three different maximal flows for the network of *Figure 15.2;* one of the alternatives to the flow of *Figure 15.4* is shown in *Figure 15.6.*) Under these circumstances we may wish to select from among the maximal flows one

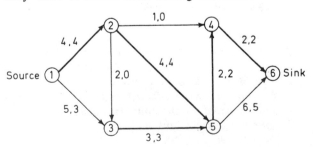

Figure 15.6

which minimises some quantitative measure, which we describe as the *flow cost*. This problem can be stated precisely as follows:

Each arc (n_i, n_j) of a network $G = (N, A)$ has a capacity $c(n_i, n_j)$ and a non-negative cost $\ell(n_i, n_j)$. It is required to construct a maximal network flow ϕ from source to sink for which the *total flow cost*

$$\ell(\phi) \;=\; \sum_{(n_i,\ n_j)\ \in\ A} \ell(n_i,\ n_j).\phi(n_i,\ n_j) \qquad\qquad (9)$$

is minimal. As an example, in a transportation network $\ell(n_i, n_j)$ might represent the cost of transporting one unit of a commodity along (n_i, n_j), in which case $\ell(\phi)$ would be the total transportation cost incurred by the flow ϕ. Again, in a traffic network, $\ell(n_i, n_j)$ might represent the *time* taken by a vehicle to traverse (n_i, n_j); in this case $\ell(\phi)$ would be the total time spent by vehicles in travelling from the source to the sink. In both cases it is evident that a minimal-cost flow is particularly desirable.

To construct a minimal-cost maximal flow we proceed as follows. Suppose that initially we have some flow ϕ, of value v say, whose cost is minimal over all flows of value v. (It is easy to obtain such a flow; for since $\ell(\phi) \geqslant 0$ for every flow ϕ, the null flow is obviously a minimal-cost flow of value zero.) Now let us construct the incremental network $G'(\phi)$ as before, except that now we also assign a cost $\ell'(n_i, n_j)$ to each arc (n_i, n_j) of $G'(\phi)$, where (cf. construction rules (1) and (2) of Section 15.4)

1. Each *normal* arc (n_i, n_j) of $G'(\phi)$ has a cost $\ell'(n_i, n_j) = \ell(n_i, n_j)$
2. Each *inverted* arc (n_j, n_i) of $G'(\phi)$ has a cost $\ell'(n_j, n_i) = \ell(n_i, n_j)$.

Suppose now that there exists some path μ from n_s to n_t on $G'(\phi)$, and let γ be the associated chain on G. If we modify the flow ϕ by increasing (decreasing) the flow in each forward (reverse) arc of γ by one unit, the *incremental cost* associated with this modification is the sum of the arc costs over forward arcs of γ minus the sum of the arc costs over reverse arcs of γ. Hence the incremental cost is given by

$$\ell(\mu) \;=\; \sum_{(n_i,\ n_j)\ \in\ \mu} \ell(n_i,\ n_j) \qquad\qquad (10)$$

the sum of the costs of the arcs of μ (called the *cost of the path* μ). It follows that a least-cost path from source to sink $G'(\phi)$ determines a best-possible chain on G for modifying the flow ϕ. Furthermore, it is possible to prove the following:

Let ϕ be a minimal-cost flow of value v, let μ be a least-cost path on $G'(\phi)$ of minimal arc capacity ϵ, and let γ be the chain on G associated with μ. Then the flow obtained by increasing (reducing) the flow in forward (reverse) arcs of γ by ϵ units is a minimal-cost flow of value v + ϵ.

The proof is unfortunately too long for inclusion here, but the interested reader will find it in some of the texts listed in the references (Busacker and Saaty, 1965; Ford and Fulkerson, 1962; Roy, 1970).

It follows that a minimal-cost maximal flow will be constructed by the maximal-flow algorithm of the previous section, provided that the computation is initiated with a minimal-cost flow, and that at each stage, the path μ chosen on $G'(\phi)$ is of minimal cost. The use of this method will be demonstrated in the next section.

15.6. Some network flow models

In this section it will be demonstrated that, by various simple extensions, our network flow methods can be applied to a wide variety of practical problems. In the first example, a transportation problem, it will be shown how to deal with multiple sources and sinks, with constraints on their inputs and outputs. The second example will demonstrate the construction and use of a network flow model to solve an assignment problem which does not involve a network in any physical sense. The last example will demonstrate the use of a network flow model for determining an optimal purchasing policy.

EXAMPLE 1: A TRANSPORTATION PROBLEM

There are m coalfields x_1, x_2, ..., x_m, each coalfield x_i being able to produce $c(x_i)$ tons per week, at a cost of $\ell(x_i)$ pounds sterling per ton. There are n power stations y_1, y_2, ..., y_n, each power station y_i requiring $c(y_i)$ tons per week. The costs $\ell(x_i, y_j)$ of transporting one ton from each coalfield x_i to each power station y_j are specified. How much coal should be produced at each coal-field, and where should it be sent, in order to meet all the demands at minimum total cost?

This problem can be considered as the problem of finding a minimal-cost maximal flow on the network of *Figure 15.7,* in which the first

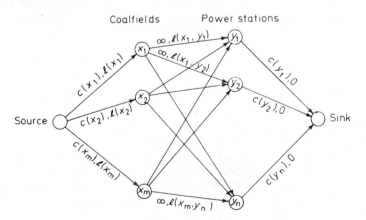

Figure 15.7

209

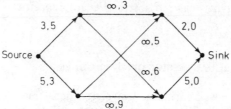

Original Network
(the numbers indicate the arc capabilities and costs)

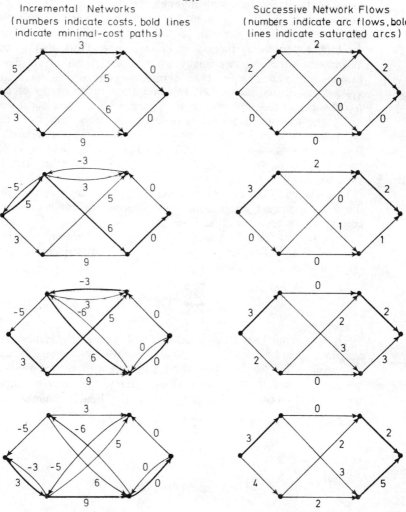

Incremental Networks
(numbers indicate costs, bold lines
indicate minimal-cost paths)

Successive Network Flows
(numbers indicate arc flows, bold
lines indicate saturated arcs)

Figure 15.8

number associated with each arc is its capacity, the second its cost. *Figure 15.8* demonstrates the application of our algorithm to such a problem, having two coalfields and two power stations. As an exercise, the reader may care to insert the arc capacities on the incremental networks.

EXAMPLE 2:   AN ASSIGNMENT PROBLEM

In a rather well-organised University department, projects are assigned to students in the following manner. First, each of the m project supervisors s_1, s_2, ..., s_m is asked to produce a list of suitable topics. There is no limit to the number of topics which he can suggest, but for each supervisor s_i there is a known upper limit $c(s_i)$ to the number of projects which he can finally undertake. A list of all the n proposed projects p_1, p_2, ..., p_n is then compiled, after which the r students u_1, u_2, ..., u_r are each asked to state their order of preference. A network of the type shown in *Figure 15.9* is then constructed, in which the number associated with each arc is its capacity. Each arc (u_i, p_j) from a student to a project also has an integral cost $\ell(u_i, p_j) = k$, $1 \leqslant k \leqslant n$, which signifies that student u_i considers project p_j to be the kth most appealing. For all other arcs the costs are zero.

It is evident that the integral maximal flows on this network define the feasible project assignments. (To obtain the project assignment corresponding to a maximal flow ϕ, we simply assign to student u_i that project p_j for which $\phi(u_i, p_j) = 1$.) Furthermore, a minimal-cost maximal flow defines an assignment giving each student u_i a project $P(u_i)$ in such a way that

$$\sum_{i=1}^{r} \ell(u_i, P(u_i))$$

is minimised.

Thus, the maximal flow algorithm will give a reasonably satisfactory assignment, and the students which it treats less favourably can perhaps be consoled by the fact that their suffering will be for the common good. However, if they are mathematicians, they can be expected to argue that to obtain a fair assignment one should minimise not

$$\sum_{i=1}^{r} \ell(u_i, P(u_i))$$

but

$$\max_{1 \leqslant i \leqslant r} \{ \ell(u_i, P(u_i)) \}$$

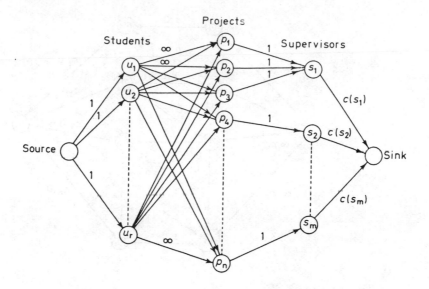

Figure 15.9

In this case the problem becomes a *bottleneck assignment problem,* of a type which arises in the maximisation of flow through production lines. This problem can be solved, by obtaining a sequence of maximal flows on networks derived from G by successively removing some of its arcs (Ford and Fulkerson, 1962).

EXAMPLE 3: A STOCK CONTROL PROBLEM

A firm buys raw material for a manufacturing process at the beginning of each month. For each of the next n months, estimates have been made of the amount c_i of the material which will be used in the ith month, and of the unit purchase price p_i at the beginning of this period. The firm can store up to s units of the material. What quantity of material should be bought at the beginning of each month, for the total expenditure over the n-month period to be minimal?

A minimal-cost maximal flow on the network of *Figure 15.10* gives an optimal purchasing policy, the flow in the arc from the source to node m_i giving the amount to be purchased at the beginning of the ith month. It will be seen that by simple modifications of the network, one can also take into account (a) a non-zero initial stock, (b) a final stock requirement, and (c) storage costs and loss of interest on expended capital.

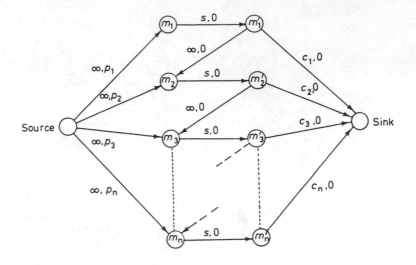

Figure 15.10

15.7.   Extensions

In constructing a network flow model of a particular problem, it is frequently found that the model is more complicated in nature than those which we have considered above. For instance, there are often flow constraints which cannot be represented simply by assigning capacities to arcs, and sometimes the cost of sending a commodity along an arc is a non-linear function of the arc flow. However, in such a case it is often possible to transform the network model into a model of the type which we have considered, such that a minimal-cost maximal flow on the latter gives the required solution. Most of the transformation methods are rather complicated, and we cannot enter into the details here. Instead, we simply give an indication of some of the types of networks for which such transformations exist:

UPPER AND LOWER BOUNDS ON ARC FLOWS

Previously we have assumed that the arc flow constraints are all of the form

$$0 \leqslant \phi(n_i, n_j) \leqslant c(n_i, n_j) \tag{11}$$

However, for some problems the arc flows in the network model G must meet constraints of the form

$$b(n_i, n_j) \leqslant \phi(n_i, n_j) \leqslant c(n_i, n_j) \tag{12}$$

where $b(n_i, n_j)$ is a specified non-negative integer. If there exist any flows which satisfy these conditions, then a feasible flow ϕ on G can always be obtained by finding a maximal flow ϕ' on a network G' derived from G, the latter having arc flow constraints of the form (1) (Berge and Ghouilha-Houri, 1965; Busacker and Saaty, 1965; Ford and Fulkerson, 1962). The same technique makes it possible to find maximal flows from source to sink on G, and also minimal-cost circulatory flows on G. A problem of this kind arises in finding an optimal assignment of railway rolling-stock, to meet a specified time-table (Roy, 1970).

NON-LINEAR ARC COSTS

If the cost of an arc flow is a convex function of the flow, for which a piece-wise linear representation can be used, it can be represented by using several 'parallel' arcs having costs and capacities of the type discussed previously. It is easy to prove that the minimal-cost maximal flow algorithm of Section 15.5 gives an optimal solution for the linearised model (Roy, 1970). Convex cost functions arise for instance in electric power transmission networks and road traffic networks (the 'cost' in the latter case representing traversal time, which is a non-linear function of traffic density).

MAXIMAL DYNAMIC FLOWS

For a network G in which each arc has both a capacity and a traversal time, it is sometimes necessary to determine the maximum amount of a commodity which can reach the sink from the source in a specified number p of time periods, and to find a shipping schedule which achieves this result. This dynamic problem on G can be converted into a static maximal flow problem on a time-expanded version $G(p)$ of G, where corresponding to each node n_i of G, $G(p)$ has $p + 1$ nodes $n_i(T)$, $T = 0, 1, ..., p$. This technique, and others are described by Ford and Fulkerson (1962).

15.8. Alternative modelling methods

First we make two simple observations:

1. The maximal flow problem defined in Section 15.4 is a particular case of the minimal-cost maximal flow problem of Section 15.5, obtainable from the latter by setting all arc costs to zero.
2. Given a network $G = (N, A)$ with source n_s and sink n_t, let $\tilde{G} = (N, \tilde{A})$ be the network obtained by adding to G an arc from n_t to n_s, this arc having an infinite capacity and a very

large negative cost. Then the minimal-cost maximal flows from n_s to n_t on G correspond to the minimal-cost flows of value zero on $\tilde{G}$.

Now let us define c_{ij} and ℓ_{ij} (for $i = 1, 2, ..., p$ and $j = 1, 2, ..., p$) by

$$c_{ij} = \begin{cases} c(n_i, n_j) & \text{if} \quad (n_i, n_j) \in \tilde{A} \\ 0 & \text{if} \quad (n_i, n_j) \notin \tilde{A} \end{cases}$$

$$\ell_{ij} = \begin{cases} (n_i, n_j) & \text{if} \quad (n_i, n_j) \in \tilde{A} \\ 0 & \text{if} \quad (n_i, n_j) \notin \tilde{A} \end{cases}$$

(13)

Then it follows from our observations above that all the problems considered previously can be posed as linear programs, of the following form: Find a set of real numbers ϕ_{ij} ($i = 1, 2, ..., p$; $j = 1, 2, ..., p$) which satisfy the constraints (cf. Equations (1) and (2)):

$$\sum_{j=1}^{p} \phi_{ij} - \sum_{j=1}^{p} \phi_{ji} = 0 \qquad (i = 1, 2, ..., p) \tag{14}$$

$$0 \leqslant \phi_{ij} \leqslant c_{ij} \qquad (i = 1, 2, ..., p; j = 1, 2, ..., p) \tag{15}$$

and which minimise the function

$$\sum_{i=1}^{p} \sum_{j=1}^{p} \ell_{ij} \, \phi_{ij} \tag{16}$$

The fact that all our problems can be cast in this form raises several interesting questions. In what respects does the usual linear programming approach to these problems differ from our own? Is it possible to relate the Simplex method of solving linear programs to the minimal-cost flow algorithm of Section 15.5? Under what conditions can a linear programming problem be represented by a network model? It is not possible to give the answers here, but the reader may care to investigate these matters.

Finally, we note that Examples 2 and 3 of Section 15.6 can be posed as 'multi-stage decision problems', and solved by Dynamic Programming (Bellman and Dreyfus, 1962). Again, a comparison of the modelling techniques and solution methods proves most interesting.

15.9.   References

BELLMAN, R.E. and DREYFUS, S.E. (1962). *Applied Dynamic Programming,* Princeton University Press; New Jersey

BERGE, C. and GHOUILA-HOURI, A. (1965). *Programming, Games and Transportation Networks,* Methuen; London

BUSACKER, R.G. and SAATY, T.L. (1965). *Finite Graphs and Networks,* McGraw-Hill; New York

FORD, L.R. and FULKERSON, D.R. (1962). *Flows in Networks,* Princeton University Press; New Jersey

HU, T.C. (1969). *Integer Programming and Network Flows,* Addison-Wesley; Reading, Massachusetts

ROY, B. (1970). *Algèbre Moderne et Théorie des Graphes,* 2, Dunod; Paris

15.10. Problems for further study

1. In transportation and communication networks there is usually an upper limit to the flow which can traverse each *node*. How would you represent restrictions of this kind, in a model with upper bounds on arc flows only?

2. *The family excursion problem:* There are p families f_1, f_2, ..., f_p, which want to go for an excursion in q cars c_1, c_2, ..., c_q. Given the number of members m_i of each family f_i, and the number of seats s_j in each car c_j, is it possible to find a seating arrangement such that no two members of the same family are in the same car?

3. Prove the following theorem:

Let ϕ be a flow of value v on a network $G = (N, A)$; then a necessary and sufficient condition for ϕ to be of minimal cost (taken over all flows of value v) is that the incremental network $G'(\phi)$ does not contain any cycles of negative cost.

For guidance, *see* Busacker and Saaty (1965) or Roy (1970).

4. Using the above theorem, prove the theorem of Section 15.5, which justifies the algorithm for constructing minimal-cost maximal flows.

5. Transportation problems of the type described in Section 15.8 are often solved by the 'stepping-stone' method – a particular form of the Simplex method of linear programming (for details, *see* for instance G.B. Dantzig, *Linear Programming and Extensions,* Princeton University Press, 1963). Solve the problem of *Figure 15.8* by this method.

Describe the stepping-stone method (in particular the nature of the initial 'basic feasible solution', and the iterative process) in terms of flows on networks.

6. Previously it has been assumed that only one substance or 'commodity' flows through a network, and that when a network has several sources and sinks it is possible to supply any sink from any source.   However, suppose that we have k commodities, that we distinguish on a network k sources s_1, s_2, ..., s_k and k sinks t_1, t_2, ..., t_k, and that we wish to send simultaneously a given number v_i of units of commodity i from s_i to t_i, for $i = 1, 2, ..., k$, there being an upper limit to the total flow along each arc.   This is an example of a *multiple-commodity flow problem,* which arises in the operation of transportation systems.   Investigate the possible methods of solving such problems (Hu, 1969; Roy, 1970).

7. What are the points of similarity and the essential differences between the types of network flows considered in this chapter and traffic flows in road networks? (*see* for instance Herman, R. (Ed.), *The Theory of Traffic Flow,* Elsevier, 1961).

16
URBAN STRUCTURE

R.H. Atkin

Department of Mathematics, University of Essex

[Prerequisites: set theory and modern algebra]

16.1. Introduction

The ideas behind mathematical modelling have developed from considerations of systems of logic (Stoll, 1961) or, in parallel with this, general
axiomatic systems such as those lying at the foundations of geometries
(Blundell, 1961). These studies pointed to the advantages to be gained
in formulating a practical system of axioms in some well-defined mathematical way; for example, the early work of Boole in the sphere of
logic and the studies of finite projective geometries. In the first case
we have a clear illustration of the use of a specific kind of *algebra* to
contain and fully represent the axiomatic system, and in the second
case we find the possibility of a *geometric* realisation of the axioms
(even if there is also an algebraic one). In other spheres we find
attempts to create a mathematical model in systems of, for example,
systems of linear differential equations or as systems of linear operators
in an Hilbert space, some of which are illustrated in this book.
   In this chapter I wish to introduce the reader to an attempt at
understanding an urban community, to giving some realistic meaning to
the words 'Urban Structure', and to do so by setting up a mathematical language which will attempt to make no concessions to simplification
or approximation; that is to say, to postulate a mathematical model
which will give us a total view of the system in the holistic sense.
Such a model, if successful, will *be* the structure, it will not 'model'
the structure in the weak sense of ignoring some of it.

16.2.   Mathematical relations in an urban community

A town consists of various finite sets of 'things'. There is a set of people, P; there is a set of human interests (activities), A; a set of buildings, B; and a set of streets, S. These may not be exhaustive, but we can always remedy that, nor need they be any more precise at this stage. For example, the set A might be described better as the union of other sets,

$$A \; = \; A_1 \; \cup \; A_2 \; \cup \; ... \; \cup \; A_k$$

where A_1 is the set of commercial retail activities,
  A_2 is the set of public services,
  A_3 is the set of private services (solicitors, doctors etc.)
  etc.

The essence of the urban community lies in the *mathematical relations* which exist between these various sets.

To be precise a set of people P, say,

$$P \; = \; \{P_1, \; P_2, \; ..., \; P_{10}\}$$

is related to a set of activities

$$A \; = \; \{A_1, \; A_2, \; ..., \; A_6\}$$

if we can give unambiguous answers to the question

$$Q: \quad \text{Is } P_i \text{ involved with } A_j?$$

for every pair of integers $(i, \, j)$; $i = 1, \, 2, \, ..., \, 10$; $j = 1, \, 2, ..., \, 6$. This give us a relation λ which possesses an *incidence matrix*

$$\Lambda \; = \; (\lambda_{ij})$$

where $\lambda_{ij} = 1$ if the answer to Q is 'yes'
      $= 0$ if the answer to Q is 'no'

We therefore obtain a typical incidence matrix Λ representing the relation

$$\begin{array}{c|c} \lambda & A \\ \hline P & \end{array}$$

of the form

$$\Lambda = \begin{bmatrix} 1 & 1 & 1 & 1 & 0 & 0 \\ 1 & 1 & 0 & 0 & 1 & 0 \\ 0 & 0 & 1 & 0 & 0 & 0 \\ 1 & 0 & 0 & 1 & 0 & 0 \\ 1 & 0 & 1 & 1 & 0 & 1 \\ 0 & 0 & 1 & 0 & 1 & 0 \\ 1 & 1 & 0 & 1 & 0 & 1 \\ 0 & 0 & 1 & 0 & 1 & 0 \\ 1 & 1 & 0 & 1 & 0 & 0 \\ 0 & 1 & 0 & 0 & 1 & 1 \end{bmatrix}$$

Notice that the answer to Q must be yes/no; 'maybe' is not allowed. Also, at this stage, the question of *ranking* the activities A_1, ..., A_6 does not arise — although it is easy to take it into account at a later stage.

Some intuitive meaning can be introduced into this relation λ by identifying the set $A = \{A_i\}$ as follows:

A_1 = the pastime of playing golf
A_2 = an interest in local people and events
A_3 = activity of gardening
A_4 = enthusiasm for motoring
A_5 = spare-time study of a foreign language
A_6 = interest in conservation and environment

Another relation μ will exist between, say, the set of buildings B and a set of activities, R, which describes the retail trade of the town. Thus if R contains elements such as

R_1 = trade in cosmetics
R_2 = dairy produce
R_3 = butchery
R_4 = fishmonger etc.

then we obtain, in a well-defined manner, an incidence matrix μ which represents the relation

$$\begin{array}{c|c} \mu & B \\ \hline R & \end{array}$$

and when $\mu_{ij} = 1$ we know that retail trade R_i is to be found located in building B_j; when $\mu_{ij} = 0$, then it is not so found.

It is clear that, for a complete description of our town, we would expect a number of relations like λ, μ, ... between various pairs of sets, like P, A, R, B, ... etc.

16.3.   The structure of a relation

A relation λ between two finite sets Y and X is a subset of the cartesian product $Y \times X$, so we can write $\lambda \subset Y \times X$. If the pair $(Y_i, X_j) \in \lambda$ then Y_i is λ-related to X_j. The relation may be represented by an *incidence matrix* $\Lambda = (\lambda_{ij})$, where

$$\lambda_{ij} = 1 \quad \text{if } (Y_i, X_j) \in \lambda$$

and

$$\lambda_{ij} = 0 \quad \text{if } (Y_i, X_j) \in \lambda$$

Each such relation λ gives rise to a *simplicial complex*, denoted by $K_Y(X; \lambda)$, and this complex (or its abstract geometrical realisation) is what we shall mean by the 'structure of λ'. The complex $K_Y(X; \lambda)$ is defined as follows.

1. $K_Y(X; \lambda)$ is a collection of simplices $\{\sigma_p; p = 0, 1, ..., N\}$.
2. Each $\sigma_p \in K$ is uniquely defined by a subset of $(p + 1)$ distinct X_i for which there is at least one $Y_k \in Y$ such that $(Y_k, X_i) \in \lambda$ for each of the $(p + 1)$ values of i.
3. The $\sigma_0{}^i$ are identified with the X_i, $i = 1, ..., n$ (n is the number of elements of X).
4. Every $(q + 1)$ subset of a σ_p $(q < p)$ is called a *q-face* of σ_p and defines a $\sigma_q \in K$ (written $\sigma_q < \sigma_p$).

The number N in (1) is called the *dimension* of K and written dim K; it denotes the largest dimension of any $\sigma_p \in K$. The set X is also called the *vertex set* of $K_Y(X; \lambda)$; we notice too that each simplex $\sigma_p \in K$ corresponds to at least one $Y_k \in Y$ by (2). When there is no danger of confusion we might speak loosely of the set Y denoting the simplices of $K_Y(Y; \lambda)$; hence the notation.

In a similar way, if we let Y be the vertex set, λ^{-1} gives us the *conjugate complex* $K_X(Y; \lambda^{-1})$ in which the X_i denote simplices. We notice here that the incidence matrix for λ^{-1} is Λ^T, the transpose of Λ.

A simple numerical example of a relation $\lambda \subset Y \times X$, is provided by that between our hypothetical people P and activities A on p.219 above. In this case $Y = P$ and $X = A$.

Considering the complex $K_P(A; \lambda)$ we notice that

$$\langle A_1\ A_2\ A_3\ A_4 \rangle \text{ is a } \sigma_3, \text{ whose name is } P_1$$
$$\langle A_1\ A_2\ A_5 \rangle \quad \text{ is a } \sigma_2, \text{ whose name is } P_2 \text{ etc.}$$

A *geometric realisation* of a complex can be obtained in a euclidean space E^H, and in general, it can be shown that $H = 2N + 1$, where $N = $ dim K. *Figure 16.1* is a realisation for $K_Y(X; \lambda)$.

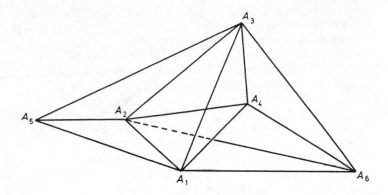

Figure 16.1

A significant concept, in a structure of this kind, is provided by that of *q-connectivity*, as follows.

Given two simplices σ_p, $\sigma_r \in K$ we shall say they are joined by a chain of connection if there exists a finite sequence of simplices

$$\sigma_{a_1}, \sigma_{a_2}, ..., \sigma_{a_h}$$

such that

1. σ_{a_1} is a face of σ_p.
2. σ_{a_h} is a face of σ_r.
3. σ_{a_i} and $\sigma_{a_{i+1}}$ share a common face (say) σ_{β_i} for $i = 1, ..., (h - 1)$.

We shall say that this chain of connection is a *q-connectivity* if q is the least of the integers

$$\{a_1, \beta_1, \beta_2, ..., \beta_{h-1}, a_h\}$$

As a special case we see that a simplex σ_p must be regarded as p-connected to itself, although it cannot be $(p + 1)$-connected to any simplex.

It is not difficult to prove that if σ_p and σ_r are q-connected then they are also $(q - 1)$-, ..., 1-, 0-connected in K.

The process of identifying the largest pieces of K which are q-connected, for all values of q from 0 to dim K constitutes a partitioning of the simplices of K at each q-level. Thus, we can introduce a relation γ_q on the simplices of K, defined by

$$(\sigma_p, \sigma_r) \in \gamma_q \text{ if and only if } \sigma_p \text{ is } q\text{-connected to } \sigma_r$$

This γ_q is reflexive, symmetric and transitive and is therefore an equivalence relation. The equivalence classes, under γ_q, are the members of the quotient set K/γ_q, and constitute a partition of K.

We denote the cardinality of K/γ_q by Q_q; this equals the number of distinct q-connected components in K.

When we analyse K by finding all the values of

$$Q_0, \; Q_1, \; ..., \; Q_N$$

where $N = \dim K$ we say that we have performed a *Q-analysis* on K.

An algorithm for finding the q-values of the shared faces of all pairs of simplices in K, and thus for deducing the Q_q-values, makes use of the incidence matrix Λ which defines K.

If the cardinalities of the sets Y and X are m, n respectively, the matrix Λ is an $(m \times n)$ matrix, with entries 0, 1. In the product $\Lambda\Lambda^T$ the number in position (i, j) is the result of the inner product of row i with row j, taken from Λ. This number therefore equals the number of 1's common to rows i and j in Λ. It is therefore equal to the value $(q + 1)$, where q is the dimension of the shared face of the simplices σ_p, σ_r represented by rows i and j. The algorithm can therefore be summarised as follows.

To find the shared face q-values between all pairs of the Y-simplices in $K_Y(X; \lambda)$

1. Form $\Lambda\Lambda^T$, an $(m \times m)$ matrix.
2. Evaluate $\Lambda\Lambda^T - \Omega$, where $\Omega = (w_{ij})$ and $w_{ij} = 1$ for $i, j = 1, 2, ..., m.$ `

The analysis for $K_X(Y; \lambda^{-1})$ follows by forming $\Lambda^T\Lambda - \Omega'$ where Ω' is an $(n \times n)$-matrix of 1's.

The complex given in *Figure 16.1* gives, by way of illustration, the following q-pattern — where we reproduce only the upper triangular 'half' of the symmetric matrix $\Lambda\Lambda^T - \Omega$, for ease of analysis, and where we write '-' for $q = -1$ (the symbol of disconnection).

P_1	P_2	P_3	P_4	P_5	P_6	P_7	P_8	P_9	P_{10}	
3	1	0	1	2	0	2	0	2	0	P_1
	2	-	0	0	0	1	0	1	1	P_2
		0	-	0	0	-	0	-	-	P_3
			1	1	-	1	-	1	-	P_4
				3	0	2	0	1	0	P_5
					1	-	1	-	0	P_6
						3	-	2	1	P_7
							1	-	0	P_8
								2	0	P_9
									2	P_{10}

The integers in the diagonal are the dimensions of the *P*-simplices and the *Q*-analysis follows from an inspection of the other entries. Thus we have

$$\dim K = 3, \text{ since } P_1, P_5, P_7 \text{ are 3-simplices}$$

and

at $q = 3$, $Q_3 = 3$, viz., $\{P_1\}$, $\{P_5\}$, $\{P_7\}$

at $q = 2$, $Q_2 = 3$, viz., $\{P_1, P_5, P_7, P_9\}$, $\{P_2\}$, $\{P_{10}\}$

at $q = 1$, $Q_1 = 2$, viz., $\{P_1, P_2, P_4, P_5, P_7, P_9, P_{10}\}$
$\{P_6, P_8\}$

at $q = 0$, $Q_0 = 2$, viz., $\{\text{all}\}$

We can thus see how the chains of connection link the people P_i in this hypothetical community, as far as the specific relation λ is concerned.

At $q = 3$, no two people are 3-connected, but P_1, P_5, P_7 all come into the same component at $q = 2$. At this level, P_2 and P_{10} are isolated, but they join the others at $q = 1$, at which level P_6 and P_8 form a separate but disconnected component. In fact, P_6 and P_8 are 1-connected through the common face (the 1-simplex $\langle A_3, A_5 \rangle$).
far as λ is concerned, P_6 and P_8 are identical people characterised by A_3 (gardening) and A_5 (foreign language). Finally, all the people fall into a single component at $q = 0$, so that there must be chains of 0-connection which joins them all together (at the worst these chains will constitute a tree with more than one branch).

16.4. Some consequences

The total collection of mathematical relations λ_i, $i = 1, ..., h$, which are used to represent the urban community gives rise to a collection of $2h$ simplicial complexes, two for each λ_i. This collection, *S*, is referred to as the *static backcloth* of that community. Expressed in abstract geometrical terms, this backcloth *S* possesses a multi-dimensional structure in a euclidean space E^{2k+1}, where $k = \max \{\dim K; K \in S\}$. It is against this backcloth (or, it is *in this geometry* of *S*) that the *dynamics* of the community must function. These dynamics will be described by changes in patterns (mathematical functions) defined on the simplices of *S*; precisely, we introduce the following:

Definition I: A *pattern* π, on a simplicial complex *K*, is a mapping

$$\pi : K \to J$$

where the image of any simplex σ_p, of K, viz., $\pi(\sigma_p) = (\sigma_p, \pi) \in J$; this J denoting a suitable arithmetic.

Incremental changes in a particular pattern π, identified as a new pattern $\delta\pi$, can now be interpreted as 'forces' or 'community pressures' acting in the complex K — in the backcloth S. But because each complex K is a graded structure there is a natural way of viewing a pattern as *graded*, writing it as

$$\pi \;=\; \pi^0 \oplus \pi^1 \oplus \ldots \oplus \pi^t \oplus \ldots \oplus \pi^N, \; N = \dim K$$

so that

$$\delta\pi \;=\; (\delta\pi)^0 \oplus (\delta\pi)^1 \oplus \ldots \oplus (\delta\pi)^N$$

and so the level, the t-level, associated with a particular $\delta\pi$ can be taken into account. We do this by introducing the following:

Definition II: If $(\delta\pi)^t \in \delta\pi$ is a non-zero pattern we say that it is associated with a *t-force* in the backcloth S and we call the ratio $(\sigma_t, (\delta\pi)^t) \div (\sigma_t, \pi^t)$, when that exists, the *intensity* of the t-force on the simplex σ_t.

Now the most important feature of these patterns $\delta\pi$ is their relationship to and dependence on the geometry of the backcloth. It is the connectivity structure of S which is dominant in this context, as the following argument shows.

Introduce an operator, f, to be called the *face*-operator which is to replace a simplex σ_p by all its $(p - 1)$ faces. Precisely, if

$$\sigma_p \;=\; \langle X_1 \; X_2 \; \ldots \; X_{p+1} \rangle$$

then

$$f(\sigma_p) \;=\; f\langle X_1 \; X_2 \; \ldots \; X_{p+1} \rangle$$

$$=\; \text{set of all faces}$$

$$=\; \overset{p+1}{\underset{i=1}{\cup}} \; \langle X_1 \; X_2 \; \ldots \; \hat{X_i} \; \ldots \; X_{p+1} \rangle$$

where $\hat{X_i}$ denotes that X_i is to be omitted.
Thus

$$f(\sigma_p) \;=\; \underset{i}{\cup} \; \sigma^i_{p-1} \; \text{where} \; \sigma^i_{p-1} \leqslant \sigma_p$$

Now there is a natural dual to this operator, which now we shall denote by Δ and call the *coface* operator, and this is related to f by defining $\Delta\pi^{p-1}$ on σ_p by

$$(\sigma_p, \ \Delta\pi^{p-1}) \ = \ (f\sigma_p, \ \pi^{p-1})$$

We notice at once that when π is defined on a p-simplex σ_p then $\Delta\pi$ is defined on a $(p + 1)$ simplex σ_{p+1}, where $\sigma_p \leqslant \sigma_{p+1}$.

If we denote all the p-simplices of a complex K by K^p then we can write

$$K \ = \ K^0 \ \cup \ K^1 \ \overset{\overset{f}{\leftarrow}}{\underset{\underset{\Delta}{\rightarrow}}{\cup}} \ ... \ K^p \ \overset{\overset{f}{\leftarrow}}{\underset{\underset{\Delta}{\rightarrow}}{\cup}} \ K^{p+1} \ \cup \ ... \ \cup \ K^N$$

and the arrows denote the directions of the operators f, Δ, remembering of course that Δ acts on patterns (on K).

For any given incremental change $\delta\pi$, of a pattern π, we can associate a new pattern $\Delta\pi$ (since it is defined by π) by writing

$$(\sigma_t, \ \delta\pi) \ = \ (f^{-1} \ \sigma_t, \ \Delta\pi) \tag{1}$$

where

$$f^{-1} \ \sigma_t \ = \ \underset{i}{\cup} \ \sigma^i_{t+1}$$

with

$$\sigma_t \ \leqslant \ \sigma^i_{t+1}$$

and

$$(\ \underset{i}{\cup} \ \sigma^i_p, \ \pi) \ = \ \underset{i}{\sum} \ (\sigma^i_p, \ \pi)$$

Thus a t-force is associated with a pattern $\Delta\pi$ defined on certain specified $(t + 1)$ simplices.

But we can now see how the geometry of S affects the role of this t-force pattern $\Delta\pi$, consequent upon an increment $\delta\pi$. If $\sigma_t \in K$ is such that $f^{-1} \sigma_t$ is the empty set, so that there are no $(t + 1)$ simplices of which σ_t is a face, *then $\Delta\pi$ is undefined.* This follows because (1) implies that the value of $\delta\pi$ on σ_t, $(\sigma_t, \delta\pi)$, is redistributed (partitioned) over the set $f^{-1} \sigma_t$ and that these components (of the partition) are attributed to $\Delta\pi$. Thus if the cardinality of $f^{-1} \sigma_t$ is $N_{t+1} = 0$ this is undefined, and so $\Delta\pi$ is undefined.

Now if we refer back to the Q-analysis of Section 16.3 we see that the numbers Q_r, $r = 0, 1, ...,$ dim K, give us a measure of how many σ_r exist which are *not* faces of some σ_{r+1}. For this reason the numbers

$$Q = \{Q_n, Q_{n-1}, ..., Q_1, Q_0\}, \quad n = \dim k$$

form a vector, called the *structure vector* $Q(K)$ of the complex K. This structure vector is therefore already an indication of the in-built geometrical *obstruction* to the existence of $\Delta\pi$, for any π. For reasons which we need not pursue in this article (but *see* Atkin (1974)) when $Q_0 = 1$, we can introduce an *obstruction vector* $\tilde{Q}$ with components

$$\{Q_n - 1, Q_{n-1} - 1, ..., Q_1 - 1, 0\}$$

which is a better indication of this effect.

Loosely speaking, a high value of the tth component of the obstruction vector $\hat{Q}(K)$ indicates that there is an inbuilt geometrical obstruction to the free change of patterns, at that t-level. Changes can only be tolerated (in the geometrical framework of S) if they take on infinite values ($N_{t+1} = 0$) — which is practically intolerable. It follows that an increase in $\hat{Q}$ (in any of its components) is a sign of increasing rigidity in the dynamics of the community.

By way of example, the town of Southend-on-Sea in Essex, UK, publishes (as do many other towns) its 'town map'. This town map shows 19 residential areas within the borough boundaries and a total of 30 civic and social amenities, which range from parks, schools (various) to the sea-side amusements and the waterworks. Although the data are somewhat minimal and rather crude, nevertheless they are sufficient to illustrate the ideas.

The Q-analysis of the relation λ gave the following results.

1. The outstanding residential area was a simplex of order 14.
2. The next two areas appeared at $q = 10$, but at that stage none of the three areas were connected; thus $Q_{10} = 3$.
3. Other areas appeared in the analysis at lower q-values, but the first sign of a lengthy chain of connection only occurred at $q = 5$; at $q = 6$ we have $Q_6 = 9$ whilst at $q = 5$ we have $Q_5 = 3$.
4. Overall the structure vector $Q(K)$ became

$$Q(K) = \{1, 1, 1, 1, 3, 5, 8, 11, 9, 3, 4, 1, 1, 1, 1\}$$

5. The obstruction vector was consequently

$$\hat{Q}(K) = \{0, 0, 0, 0, 2, 4, 7, 10, 8, 2, 3, 0, 0, 0, 0\}$$

6. This demonstrated that any pattern π_t for which $t > 4$ would suffer obstruction to change, with a decided increase in this at $t = 6$. We could immediately say that, for example, if the population density pattern is rigid then it must correspond to a π_t with $t > 4$. Generally it was borne out in fact that the

obstruction vector for the town, based on this particular data, corresponded to well-established rigidities in a number of the basic patterns one would normally consider.

Thus we can see that the interaction between human activities and the spatial distribution in a typical town structure is something which must be played out against a backcloth of a base complex with all its connectivities. The consequences of this for planning and urban development are clear for all to see.

A fascinating philosophical consequence of this point of view is that we can no longer assume that this particular human drama can be adequately fitted into the physicist's four-dimensional world of conventional space-time. The dimensionality of this human world of interaction is determined by the complexes $K(X)$ and $K(Y)$ which are used to describe the backcloth.

Let us take another simplified example in order to illustrate some of the ideas.

Take the hypothetical S containing the relation $\lambda \subset P \times A$ already discussed, with Q-analysis of $K_p(A; \lambda)$ in Section 16.3. Take a pattern π on this particular complex where

$$\pi(\sigma_p) = (\sigma_p, \pi) = \text{total expenditure by the set } P, \text{ on } \sigma_p, \text{ per annum}$$

so that $(\sigma_p, \pi) = \pounds x$ and x is a non-negative integer. If we begin by knowing the personal expenditures, say x_{ij} for each P_i on A_j ($i = 1, ..., 10; j = 1, ..., 6$) then we induce a value for π on each σ_p as follows.

$$x_1 = x_{11} + x_{12} + x_{13} + x_{14} \text{ since } P_1 = \langle A_1 \ A_2 \ A_3 \ A_4 \rangle$$

$$x_2 = x_{21} + x_{22} + x_{25} \qquad \text{since } P_2 = \langle A_1 \ A_2 \ A_5 \rangle$$

etc.

Now, for example, x_1 induces the value $x_{11} + x_{12}$ on the face $\langle A_1 \ A_2 \rangle$ of P_1, and so on. Hence we find the total pattern π and, for example,

$$(\langle A_1 \ A_2 \ A_4 \rangle, \pi) = (x_{11} + x_{12} + x_{14}) + (x_{71} + x_{72} + x_{74})$$

which expresses the fact that P_1 and P_7 share the 2-face $\langle A_1 \ A_2 \ A_4 \rangle$.

This pattern $\pi : K \to \pounds J$ is an essential part of the dynamics of this community and it is inextricably involved with the geometry of the backcloth S.

For example, if S undergoes a change by the removal of A_1 (ploughing up the golf course), then the dimensions of the simplices P_1, P_2, P_4, P_5, P_7, P_9 all decrease by unity. This is experienced by these people as a community force acting upon them. This force is directly expressed by a *change* in π, since removing A_1 results in making $x_{i1} = 0$ for all i. Thus π becomes $\pi' = \pi + \delta\pi$ and $\delta\pi$ takes negative values; for example, on the 3-simplex $\langle A_1 \ A_2 \ A_3 \ A_4 \rangle$, $\delta\pi = -\pi$, since this simplex has disappeared from S_1; on the 1-simplex $\langle A_2 \ A_4 \rangle$, $\delta\pi = 0$, since $\pi' = \pi$. Wherever $\delta\pi$ is negative we can interpret it as a 'force of repulsion' in the structure — its value on $\langle A_1 \ A_2 \ A_3 \ A_4 \rangle$ indicating the evacuation of that tetrahedron. The intensity of this 3-force of repulsion (associated with the simplex $\langle A_1 \ A_2 \ A_3 \ A_4 \rangle$) will be the ratio (value of $\delta\pi$) $\div$ (value of π), or -1.

But we are now faced with the possibility of a 'law of inertia' for a social community in that we know from our own experience that our expenditure never seems to decrease. Of course, the reasons behind this, when it is true, are varied, but some of the truth might well be contained in the following tentative postulate.

Suppose that, when A_1 is removed from S, the pattern π redistributes itself over the remaining simplices so that the total expenditure remains constant. In other words, if I am P_1 and must give up my weekend golf, then for example, I spend more on my garden to compensate. Naturally enough I experience this change as a *social force* (in the structure) — it is not an illusion.

This will result in the change $\pi \rightarrow \pi' = \pi + \delta\pi$, where $\delta\pi$ is now a pattern which need not total only negative (or zero) values on the simplices of S. We shall consequently expect to find t-forces of attraction ($\delta\pi$ positive) as well as forces of repulsion ($\delta\pi$ negative) in various pieces of the geometry of S. This situation can also arise in circumstances which do not involve any drastic changes in the static backcloth of S. The overall result is that, in a general description of a community involving not only people and activities but also buildings, streets and traffic control, we shall expect to have a variety of patterns $\{\pi_i\}$, i in some index set I. Changes in these patterns, giving patterns $\{\Delta\pi_i\}$, will be described in terms of t-forces of attraction/repulsion in the structure S. If there is to be anything in the nature of scientific 'laws' for these dynamics then we would expect them to be

functional relations among the $\Delta\pi_i$, $i \in I$

The simple example above shows the possibility of one such 'law' — the law of conservation of expenditure, a sort of law-of-inertia, expressible in the form

$$\sum_p (\sigma_p, \delta\pi) = \sum (\sigma_p, \pi) - \sum (\sigma_p, \pi') = 0$$

with $\sigma_p \in S$. Under these circumstances there must be a sort of balance in which, overall, the forces of attraction cancel the forces of repulsion.

Firstly, we can notice that, referring once more to the relation λ in Section 16.3, the people P_i (who might be replaced, in a more general context, by buildings or streets or social interests or political beliefs, etc.) can be characterised in a certain way by the geometrical connections they exhibit (in this abstract geometrical space S). To do this we introduce a number for each P_i which we shall call his *eccentricity*, Ecc (P_i). This is to measure the relative extent to which he is disconnected from his fellows.

We introduce the notions of

1. The top-q value of P_i, viz., $\hat{q} = \dim P_i$ in K.
2. The bottom-q value of P_i, viz., $\check{q}$ = the largest q-value at which P_i becomes connected to any distinct P_j.

Then we define the eccentricity of P_i as

$$\text{Ecc } (P_i) = \frac{\hat{q} - \check{q}}{\check{q} + 1}$$

Thus Ecc $(P_i) = \infty$ only when $q = -1$, that is when P_i is totally disconnected from everyone else.

From the table in Section 16.3 we see that, for example,

$$\text{Ecc } (P_1) = \frac{3 - 2}{3} = \frac{1}{3}$$

whereas

$$\text{Ecc } (P_8) = \frac{1 - 0}{0 + 1} = 1 \text{ and Ecc } (P_4) = 0$$

Thus we would say that P_8 is more eccentric than is P_1, and P_4 is, in this structure, of zero eccentricity— his top-q value equals his bottom-q value.

Generally, it would appear that each person in a community is constantly engaged in a struggle to strike a suitable value for his eccentricity Ecc (P); he wishes to be neither too uniform nor too extreme — or he might be the sort of person who strives for Ecc $(P) \to 0$ (or ∞?).

These indications of the *method* of analysis, the language of connectivity, must not be confused with the rather trivial specific illustrations. The significance of this language lies in the fact that it offers to each of us a ready 'do-it-yourself-kit' for building a model; it does not offer the specific model. In this sense it is an illustration of the 'meta-model'. Consequently, it cannot be 'wrong' at the model level, it can only be badly applied.

16.5. References

The work which has already been done in this field, and the implications for the future, can be studied by referring to the following
references.

ATKIN, R.H. (1972). 'From Cohomology in Physics to q-connectivity
in Social Science', *Int. J. Man-Machine Studies,* **4**

ATKIN, R.H. (1974). 'An Algebra for Patterns on a Complex', *Int. J.
Man-Machine Studies*, **6**

ATKIN, R.H. (1974). *Mathematical Structure in Human Affairs,* Heine-
mann Educational Books Ltd.; London

ATKIN, R.H. *et al.* (1972). Research Report I, Urban Structure Project,
University of Essex

BLUNDELL, L.M. (1961). *A Modern View of Geometry,* W.H. Freeman;
San Francisco, California

STOLL, R.S. (1961). *Sets, Logic and Axiomatic Theories,* W.H. Freeman;
San Francisco, California

16.6. Problems for further study

1. Take a real situation familiar to yourself (the home, the school,
the High Street etc.) and identify two finite sets X and Y in it. (In
the High Street, X might be the set of individual shops and Y might
be the set of different kinds of things they sell.)   Identify a relation
$\lambda \subset Y \times X$ and evaluate the two complexes $K_Y(X; \lambda)$, $K_X(Y; \lambda^{-1})$.
Find the structure vector Q and the associated obstruction vector $\hat{Q}$.
Find also the eccentricities of each Y_i and each X_j.

2. Identify patterns on your choice of (1), for example, counting
people, traffic, sales, rankings (likes and dislikes on a scale 0-10).
Consider the observed changes $\delta \pi$ in these patterns and interpret them
in terms of social forces in the structure.

17
STRUCTURAL STABILITY OF MATHEMATICAL MODELS: THE ROLE OF THE CATASTROPHE METHOD

D.R.J. Chillingworth
Department of Mathematics, University of Southampton

[Prerequisites: none]

17.1. Introduction

General ideas of *stability* are fundamental to our perception of the universe. We can form mental images only of those phenomena and structures which recognisably persist or repeat themselves in time; without some quality of stability there could be no consciousness — and indeed no biological organism.

Although this may be a sound theoretical principle, it is clearly impossible to put it into practice without first formulating a mathematical language and framework within which general ideas of stability of models can be discussed. Scarcely any such language was available, however, until the 1960s with the remarkable work of the French mathematician René Thom. In his extraordinary and original book (Thom, 1972), concerned with the whole philosophy of stability and with its relevance in particular to the study of the structure and development of living systems, Thom laid the foundations of a general theory of stability applicable to mathematical models of the most diverse kinds.

It is a basic theme in Thom's philosophy that *qualitative* understanding of phenomena is not (as is perhaps too often assumed in present-day scientific disciplines) by any means inferior to *quantitative* understanding — indeed, it is fundamental, and often far more significant. It would be futile to try to predict the precise number of leaves a tree will have: it is more important to be able to recognise the difference between an oak and an ash by their qualitative structural features. Thus in many contexts (biological, sociological and psychological in particular) our mathematics should in the first instance have a strongly

231

qualitative flavour; we can then perhaps impose quantitative details
once we have constructed a satisfactory qualitative model. It is also
worth noting that quantitative methods are often used in order to give
what are in effect qualitative answers to practical problems, such as
whether a certain bridge will collapse in high winds, or whether all
planets in the solar system will indefinitely remain in roughly elliptical
orbits.

We can approach a general definition of stability for a mathematical
model by saying that a model is *structurally stable* if sufficiently small
changes in the construction of the model itself will produce behaviour
which is in some sense qualitatively similar to the behaviour of the
original model, although of course in any given case we must say precisely what we mean by 'sufficiently small' and 'qualitatively similar'.
This then leads naturally to the notion of *breakdown of stability,* and
it is here that Thom's ideas have the most far-reaching implications.

Clearly it would be invaluable to have (a) an explicit set of criteria
for assessing the structural stability or otherwise of any given mathematical model, as well as (b) a complete description of all the ways
in which stability could break down. However, it is equally clear
that these utopian aims could never be achieved without first specify-
ing the type of model, and even for the most elementary models the
programme seems wildly ambitious. Nevertheless, as Thom demonstrated,
and as we shall see below, progress can be made.

One of the simplest examples of stability which we encounter in
mathematics is the notion of *stable equilibrium* for a dynamical system.
It is important not only in mechanics, but also in the study of more
general kinds of dynamical system – economic, biological, electrical,
and so on. There are a large number of mathematical techniques
available for analysing the way a system behaves when it is perturbed
slightly from a given equilibrium state, and thus for describing whether
an equilibrium is 'highly stable' or 'only just stable', or whether it is
stable under perturbations in some directions but not in others. Here
we shall not be dealing with this kind of question, however, but will
be interested in the following wider problem involving a more general
type of stability.

Suppose we have some kind of system in equilibrium, but the system
itself is slowly changing as various important parameters change. What
happens to the equilibrium? It is natural to expect that the equilibrium
also changes slowly. It might happen, though that as the parameters
change, the equilibrium becomes less and less stable, until for a particular value or values of the parameters the equilibrium actually becomes
unstable or even disappears altogether. As a result, the behaviour of
the parametrised system would undergo a sudden switch from one equilibrium state to another, which might be quite far away from the original one. The mathematical analysis and possible prediction of abrupt
changes of this kind is naturally going to be of great interest in any
context where a system with continuous input produces a sudden discontinuity in its output, such as the collapse of an economy or of a

building, or the appearance of sharply-defined geometrical structures in the development of a piece of biological tissue.

Little was known about the general nature of such discontinuous changes until Thom showed that for a certain fairly common kind of dynamical system the various ways in which discontinuous jumps in output occur in practice can all be described geometrically, falling into a small number of classifiable types. These types he called the *elementary catastrophes.*

The aim of this chapter is to give an accurate though not too technical formulation of Thom's theorem, and to suggest ways in which by understanding these important results and familiarising ourselves with the geometry of the elementary catastrophes we may gain new insights into constructing mathematical models for real-life situations that involve sudden 'catastrophic' discontinuities.

This is merely a first step. There is not room here to begin to describe Thom's general theory of morphologies, with its applications to natural philosophy, linguistics, and even the study of thought. For this we can only refer to Thom himself (Thom, 1972, 1974).

17.2. Systems governed by a smooth function

We must begin by making more precise some of the rather loose ideas introduced above.

Given a dynamical system (by this we mean some process or other which evolves continuously with time) we can model its behaviour mathematically by picking out a number of observable and measurable quantities $x_1, x_2, ..., x_n$, representing the state of the system at time t by the point $x(t) = (x_1(t), x_2(t), ..., x_n(t))$ in euclidean n-space, and then looking at the kind of path that the point $x(t)$ traces out as t changes. If $x(t)$ remains in one position for all time t then the system is in equilibrium as far as the quantities x_i are concerned; if $x(t)$ repeatedly traces out a closed loop then the system is in some sense behaving in a periodic way, and so on.

Very often the information which we are given about the evolution of the system (via physical, economic or other laws) is in the form of a set of *differential equations,* telling us how the point $x(t)$ varies 'infinitesimally', i.e. telling us what $\dot{x}(t) = \mathrm{d}/\mathrm{d}t\, x(t)$ is and perhaps also what the higher derivatives $\ddot{x}(t)$ etc. are at each position and time. In this case the equilibrium points are precisely those for which $\dot{x}(t) = 0$ for all t.

Whether or not the system is given in terms of differential equations, it may happen that the behaviour can be adequately described by reference to the evolution with time of just one particular real-valued function of $x(t)$. This is the kind of system we shall be considering from now on. We first make the formal definition, and then look at some elementary examples.

Definition   A dynamical system is said to be *governed by a smooth function V* if V is a smooth (i.e. infinitely differentiable) function on the space of variables $(x_1, ..., x_n)$ such that

1. The equilibria of the system are precisely the critical points of V (the points where

$$\frac{\partial V}{\partial x_1} = \frac{\partial V}{\partial x_2} = ... = \frac{\partial V}{\partial x_n} = 0$$

2. Away from the equilibria the function $V(x(t))$ decreases as t increases.

EXAMPLE 1

Consider a biological cell whose behaviour is being influenced by the concentration $x(t)$ of a certain chemical substance in the cell at each instant of time t.   As a simple hypothesis let us suppose that the substance is entering the cell from outside at a constant rate S and is being destroyed inside the cell at a rate proportional to its concentration.   This means that $x(t)$ satisfies the differential equation

$$\dot{x} = -ax + S$$

If there are two cells of exactly the same type, both influenced by the same substance but not interacting with each other in any way, then the concentrations $x_1(t)$, $x_2(t)$ in each cell satisfy the equations

$$\dot{x}_1 = -ax_1 + S$$
$$\dot{x}_2 = -ax_2 + S$$

Now we complicate the model by assuming that the cells do interact at a rate which is proportional to the difference in concentrations between the two.   This converts the equations into the following system:

$$\dot{x}_1 = -ax_1 + S + D(x_1 - x_2)$$
$$\dot{x}_2 = -ax_2 + S + D(x_2 - x_1)$$

Defining the function V by

$$V(x_1, x_2) \equiv -S(x_1 + x_2) + \frac{a}{2}(x_1{}^2 + x_2{}^2) - \frac{D}{2}(x_1 - x_2)^2$$

we see that

$$\dot{x}_1 = -\frac{\partial V}{\partial x_1}$$

$$\dot{x}_2 = -\frac{\partial V}{\partial x_2}$$

or, simply,

$$\dot{x} = -\,\text{grad}\;V$$

This is an example of a *gradient system,* which is clearly a particular case of a system governed by the potential function V. Here $n = 2$.

For a discussion of this and some other dynamical systems related to biology see Chapters 1 and 2 of Rosen (1972) as well as the books by Rosen (1970) and Pavlidis (1973).

EXAMPLE 2

Again suppose $n = 2$, and let $V(x_1, x_2)$ be any smooth function on the (x_1, x_2)-plane. The graph of V is a 2-dimensional surface which can be thought of as a 'landscape', with V measuring height above sea-level. Now imagine a ball-bearing placed somewhere on the landscape and allowed to roll under the influence of gravity (acting in the direction of negative V). The net force on the ball-bearing will be in the direction of the line of greatest slope of the landscape, so the only points of equilibrium will be the critical points of the function V. Also, if we assume the existence of enough (idealised) friction to stop the ball-bearing from 'overshooting the mark', the motion away from equilibria will be always downhill — either approaching a critical point of V or ever decreasing towards $V = -\infty$. Therefore this system is an example of a system governed by a smooth function V.

Note that Example (2) itself provides a model for any system governed by a smooth function: the x-coordinates of the position of the ball-bearing correspond to the state of the system at a given time t, and the path traced out by the ball-bearing is a model for the evolution of the system from a given initial state. The analogy is closest when $n = 2$, of course; in general we have to think in terms of an n-dimensional 'landscape'.

In the use of these models in practice it may not always be necessary to insist on the presence of the large amount of damping required to prevent all 'overshooting' or small oscillations as the state homes in on an equilibrium. Provided these transient oscillations die away quickly we can usually ignore them, and admit that the landscape of V gives a good enough model for describing the behaviour of the system. In particular, the *potential energy,* rather than the total energy, may serve for describing the evolution of certain mechanical and other systems even though they might not in fact be 'governed' by their potential energy function in the strict sense defined above.

EXAMPLE 3[*]

In a simple model for the behaviour of a shallow arch under the application of a vertical load, the equation relating the downward load q to the rise of the arch in equilibrium (as a proportion p of its original ($q = 0$) value) is

$$1 - kq = p(1 - \frac{1}{m}(1 - p^2))$$

where k, m are positive constants. Calculating the work done in decreasing the rise from the original value a to pa under a constant downward load Q we find that the potential energy is

$$V = Qa(p - 1) - \frac{a}{k}\left(p - \frac{1}{2}\left(1 - \frac{1}{m}\right)p^2 - \frac{1}{4m}p^4\right) + \text{constant}$$

$$= \frac{a}{k}\left(\frac{1}{4m}p^4 + \frac{1}{2}\left(1 - \frac{1}{m}\right)p^2 - (1 - Qk)p\right) + \text{constant}.$$

We can regard the motion of the arch as a dynamical system governed by V, since in practice there should be large damping of oscillations.

It is straightforward to verify that when $m \geqslant 1$ there is only one (real) equilibrium value of p, whatever Q may be. On the other hand, and more interestingly, when $m < 1$ there are certain values of Q for which there are three equilibrium values of p, two stable (one positive value of p, and one negative — corresponding to a buckled arch) and one unstable.

Examples of this kind are very familiar in the theory of elastic stability of structures. See for instance Sewell (1976) or the wealth of examples in Thompson and Hunt (1973).

17.3.   Critical points and structural stability

If a system is governed by a smooth function V then clearly the shape of the graph of V determines the way the system behaves. What can we expect the graph to look like in general, especially in the vicinity of its critical points? When $n = 1$ there can be maxima and minima and also points of inflexion and perhaps even whole regions where the graph is horizontal. Now there are strong reasons why, when constructing mathematical models of real-life systems, we should tend to reject (or at least treat with suspicion) any function

[*]I am grateful to M. Goresky for drawing attention to this example in Timoshenko (1936)

$V(x)$ which has any critical points that are neither maxima nor minima. Such a critical point x_0 would satisfy $V'(x_0) = 0$ and $V''(x_0) = 0$ simultaneously, which means that the graph of V' would actually be tangent to the x-axis at x_0. By changing V slightly it is always possible to arrange that the graph is no longer anywhere tangent to the x-axis, and therefore that all critical points of the new V are genuine maxima or minima. Moreover, once we have done this, we see that all *sufficiently small* changes that we might now make in the new V will preserve this property of having only critical points which are maxima or minima (provided our small changes also have sufficiently small first and second derivatives). This means that the new V is a more satisfactory model than the old for two reasons:

1. Any real system will in practice be undergoing small perturbations at all times through 'background noise', and so any particular properties of V which can be destroyed by arbitrarily small changes in V and its derivatives are not necessarily accurate reflections of the properties of the system.
2. Measurements can be taken only to within a non-zero margin of error, so the properties of V which can be destroyed by changes within that margin may have no relevance at all to the real system.

The above arguments underline the importance of ensuring that *any mathematical model of a real system should have some kind of overall stability, corresponding to the inherent stability of observable phenomena in the real world.* Following Thom, we call this the hypothesis of *structural stability.* It is a basic principle of mathematical modelling, and plays a very important part in the whole of Thom's philosophy and mathematics.

For functions of two variables x_1, x_2, the picture is a little more complicated. It can be shown in this case that the only types of critical points which are not removable by arbitrarily small perturbations of the function are maxima, minima and saddle-points (maximum in one direction, minimum in another); *see Figure 17.1.* Similarly, for

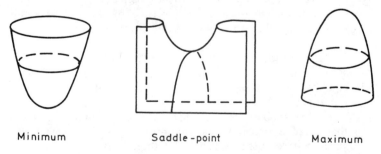

Minimum   Saddle-point   Maximum

Figure 17.1

functions of n variables x_1, x_2, ..., x_n, the important critical points are maxima, minima and $(n - 1)$ types of n-dimensional saddle-point (with maximum in r dimensions, minimum in $n-r$ dimensions, for each $r = 1$, 2, ..., $(n - 1)$). Thus if we accept the hypothesis of structural stability we see that the functions which are of particular interest to us are those whose only critical points are maxima, minima or saddle-points, and it is the behaviour of these which we must study if we want to understand the evolution of real-life systems governed by smooth functions.

Clearly the only such critical points which correspond to stable equilibria are the minima; in the other cases most (if not all) disturbances from equilibrium will cause the system to fall further away from the equilibrium state. Since we are assuming that in real life the system is all the time subject to background noise and small disturbances of one kind or another, it will be only the *minima* which in practice correspond to *observable states of equilibrium.*

17.4. Equilibria for systems controlled by k parameters

At this point we will complicate the situation by supposing that the entire system depends on a number of parameters c_1, c_2, ..., c_k. For example, it may be that we have physical control over the system and can alter some of its properties (expressed by the c_i's) as we wish, as would be the case with varying Q or perhaps m in Example (3), or it may be that we are really interested in the simultaneous behaviour of a whole family of systems, as in Example (6) below. If we suppose that for each value of $c = (c_1$, c_2, ..., $c_k)$ the system is governed by a smooth function V_c then the equilibria of the system will clearly vary with c. Now we make the following two assumptions about the way things depend on the parameters $(c_1$, c_2, ..., $c_k)$:

1. $V_{(c_1,c_2,...,c_k)}(x_1$, x_2, ..., $x_n)$ is a smooth function of x_1, x_2, ..., x_n and of c_1, c_2, ..., c_k.
2. For each c the system governed by V_c attains equilibrium so rapidly that we can consider it as essentially instantaneous.

In other words, as we vary c (thinking of it as a 'slow' variation) the system can be regarded as at all times having arrived at a state of equilibrium corresponding to a critical point – in fact a minimum – of the function V_c.

The next observation is that for some values of c it may happen that there is precisely one minimum of V_c, i.e. one stable equilibrium state, and for other values of c there may be several possible minima, with the result that a conflict is set up between different equilibrium states. We can illustrate this vividly by the following example.

EXAMPLE 4 (The Zeeman Catastrophe Machine (Zeeman, 1972))

The machine consists of a rotating arm with two elastic bands attached to the free end, mounted on a board so that the rotation is constrained to lie in a plane. One of the elastic bands has its other end anchored at a fixed point, and the second band can be stretched (if necessary) from the end of the rotating arm to any chosen point (the *control point*) on the board. *See Figure 17.2.* The control point can be given coordinates (c_1, c_2) with respect to suitable axes in the board. For some points (c_1, c_2) the arm finds only one position of stable equilibrium, but for others there are two stable equilibria. Since the equilibria of the system are determined by the tensions in the elastic bands, Hooke's Law suggests that the system is very likely to be one which is governed by a potential energy function, stable equilibria occurring where the potential energy of the stretched elastic is a minimum.

Calculations of the equations of motion show that this is correct. For some points $c = (c_1, c_2)$ the potential function (regarded as a function of θ for, say, $0 < \theta < 2\pi$) has one minimum, for others it has two. Clearly, if there are two minima there must be a maximum

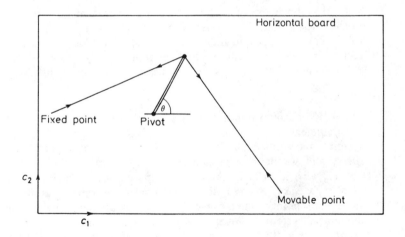

Figure 17.2   The Zeeman Catastrophe Machine

in between, and it is indeed the case that when the control point c is chosen so that the machine has two stable equilibria there is always an unstable equilibrium between them.

(There is a detailed description of the machine in Poston and Woodcock (1973). To understand the behaviour of the machine, however, there is no substitute for making one yourself.)

As the control point c is slowly moved around, the equilibrium state of the machine usually varies slowly as well, but sometimes gives a sudden and possibly violent jump from one position to another. The explanation in terms of the potential function is that as c varies it is possible for a minimum and a maximum of V_c to coalesce and disappear. If the machine were in a state corresponding to the minimum which disappears it would have no option but to fly to the other minimum. *See Figure 17.3.* This is an example of what Thom calls a *catastrophe*. It could indeed correspond to a real catastrophe if the machine were very large or if it were a model for, say, an economic system.

Figure 17.3

In the above example it is obviously important to try to find a precise description of the set of points (c_1, c_2) at which catastrophes can occur. This, though, is simply a special case of the general problem for a system with n variables governed by a smooth function with k control parameters, which we now formulate precisely in the following way:

Let V be any smooth function of n variables $(x_1, x_2, ..., x_n)$ with k parameters $(c_1, c_2, ..., c_k)$. Regarding V as governing a dynamical system as described above, we call $(c_1, c_2, ..., c_k)$-space the *control space*, and we define the *catastrophe set K* to be the set of points $c = (c_1, c_2, ..., c_k)$ in the control space for which V_c (as a function of $x = (x_1, x_2, ..., x_n)$) has some coalescent critical points. (This is the set of points where we can expect catastrophic behaviour in the dynamical system.) When $n = 1$ this means that K is the set of points c such that V_c' and V_c'' both vanish simultaneously for some x. When $n > 1$ the description is more complicated: here K is the set of points c such that all the partial derivatives

$$\frac{\partial V_c}{\partial x_1}, \frac{\partial V_c}{\partial x_2}, ..., \frac{\partial V_c}{\partial x_n}$$

and also the determinant of second derivatives

$$\begin{vmatrix} \dfrac{\partial^2 V_c}{\partial x_1 \partial x_2} & \cdots & \dfrac{\partial^2 V_c}{\partial x_1 \partial x_n} \\ \\ \dfrac{\partial^2 V_c}{\partial x_n \partial x_1} & \cdots & \dfrac{\partial^2 V_c}{\partial x_n \partial x_n} \end{vmatrix}$$

vanish simultaneously for some $(x_1, x_2, ..., x_n)$. An equation for K in terms of the $(c_1, c_2, ..., c_k)$ can in principle be found by eliminating $(x_1, x_2, ..., x_n)$ from these $n + 1$ equations. In practice, though, it may not be easy to do.

EXAMPLE 5

Let $n = 1$, $k = 2$ and $V_{(c_1, c_2)}(x) = x^3 + c_1 x^2 + c_2 x$. The critical points of $V_{(c_1, c_2)}$ occur where $V'_{(c_1, c_2)}(x) = 0$, i.e.

$$3x^2 + 2c_1 x + c_2 = 0$$

Critical points coalesce where $V''_{(c_1, c_2)}(x) = 0$, i.e.

$$6x + 2c_1 = 0$$

Eliminating x from these equations we get

$$\frac{1}{3} c_1{}^2 - \frac{2}{3} c_1{}^2 + c_2 = 0$$

i.e.

$$c_2 = \frac{1}{3} c_1{}^2$$

which is the equation of a parabola in the (c_1, c_2)-plane. This is the catastrophe set K for the given V. When

$$c_2 < \frac{1}{3} c_1{}^2$$

there is one minimum (and one maximum) for $V_{(c_1, c_2)}$; when

$$c_2 > \frac{1}{3} c_1{}^2$$

there are no critical points. As the point (c_1, c_2) moves across K with c_2 decreasing the minimum coalesces with the maximum and disappears. A system governed by $V_{(c_1, c_2)}$, and in equilibrium at the minimum of $V_{(c_1, c_2)}$, would exhibit a catastrophic jump (towards $V = -\infty$, i.e. $x = -\infty$) as this happened.

We are interested in the behaviour of a dynamical system governed by V, and so it is important to know how catastrophes occur in general, i.e. to know the local behaviour near each point c of the catastrophe set K. At this stage we remember that we are trying to make models of real-life systems, and we once again invoke the principle (discussed in section 17.3) that the only mathematical structures important to us are those which are not destroyed by arbitrarily small perturbations of the whole model. Let us say that a catastrophe set K is *stable* if its geometrical structure is not destroyed by sufficiently small changes in the function V. (By this we mean technically that if $W_c(x)$ is sufficiently close to $V_c(x)$, and all partial derivatives of W with respect to the c's and x's are sufficiently close to the corresponding derivatives of V, then there is a smooth change of coordinates in the control space which will take the catastrophe set of W to that of V.) We can then formulate our general problem as follows:

Problem: What possible local geometrical forms can stable catastrophe sets take? What is the behaviour of the associated catastrophes?

The theorem of Thom answers these questions for cases with $k \leqslant 4$ parameters.

17.5. The delay situation and the Maxwell situation

Before we go on to give a statement of Thom's theorem it is important to note that there are two particular ways in which the geometry of the catastrophe set can be related to the actual occurrence of catastrophes in the system being modelled. In the catastrophe machine example we saw that the system would remain in a stable equilibrium corresponding to a given minimum of the function V and would move slowly as the minimum moved, until the minimum disappeared (as c passed through the catastrophe set K) and a catastrophe occurred. This is how many systems, especially mechanical systems, would behave. However, some systems (in biology, for example) behave differently in that they always adopt the equilibrium state corresponding to the *lowest* minimum available. This means that the catastrophe set K will itself play no physical role, but that associated to K there is another set S of points c for which V_c has two or more minima of precisely equal heights, and catastrophes are liable to occur as the control point passes through S. This set S is called the *shock wave* associated to K, since the theory of shock waves in a piston involves just such a phenomenon. When we are interpreting facts about catastrophe sets in terms of observable 'jump' discontinuities, we therefore have to know which of the following two alternatives (if either!) apply to our system:

1. *Delay situation* The system remains in equilibrium corresponding to a given minimum for as long as possible, and

jumps to another minimum only when the first
one disappears.
2. *Maxwell situation** The system chooses an equilibrium corresponding to the lowest minimum available.

The selection of alternative depends on the physics or biology etc.,
and is *not* derived from the mathematics. It is an extra component
to be added to the mathematical model.

Once we know which situation we are in we know whether it is the
geometry of the catastrophe set K or of the shock wave S which we
should be studying in order to understand the way catastrophes occur.

To illustrate how the structure of K or S is important in physical
terms, we give an idealised example from biology showing how the
geometry of catastrophe sets and shock waves can possibly be observed
in the geometry of biological forms. This is the kind of example considered extensively by Thom in his book (Thom, 1972). One of the
most significant consequences of the theory of catastrophes is that the
beginning of a mathematical theory of the development of biological
form (*morphogenesis*) is now in principle feasible.

EXAMPLE 6

Suppose we have a piece of homogeneous biological tissue, and we
are interested in a particular property of each cell which (at least
theoretically) can be quantified. We will call it the *state* of the cell.
To be specific, let us think of this property as *colour,* quantified in
each cell by the concentration of a certain pigment (cf. Example (1)).
Now we make the assumption that the concentration in each cell is
determined by the interactions of a (possibly extremely large) number
n of chemicals with concentrations x_1, x_2, ..., x_n, and moreover that
the evolution of the x_i in each cell C is a dynamical system governed
by a smooth function V_C. Next we assume

1. The number of cells in the tissue is so large that each cell C
   may be regarded as a point in space with coordinates (c_1, c_2, c_3),
   so V_C becomes $V_{(c_1,c_2,c_3)}$.
2. $V_{(c_1,c_2,c_3)} (x_1, x_2, ..., x_n)$ is a smooth function of the x's and
   the c's.
3. In each cell $x = (x_1, x_2, ..., x_n)$ attains equilibrium instantaneously.

Having made all these assumptions we now observe that we are describing the state of the cells in the tissue by means of a mathematical
model which is precisely of the kind (dynamical system governed by a
function with parameters) that we are concerned with in this chapter
(*see* Section 17.4). Here it is important to grasp that the *control*

*After the physicist J.C. Maxwell who encountered this phenomenon in his work
on phase transitions

space in this example is the region of 3-space corresponding to the *tissue itself*: this is the essence of (1) above. On the other hand, the variable-space is of dimension *n* (maybe *n* = 5 000) corresponding to the many chemicals involved. The catastrophe set *K* and the shock wave set *S,* as subsets of the control space, correspond to the set of points (cells) *C* in the tissue where the configuration of possible equilibria of V_C changes – in other words, where the state may change abruptly. Now the precise way in which the state changes will depend on other physical and biological factors which are not part of our model, but if we assume either the delay situation or the Maxwell situation then it is clear that *the geometric forms of the catastrophe set* K *and shock wave set* S *are strongly related to the geometric form of observable discontinuities in the state of the tissue.* Hence we can expect that a general understanding of the geometrical structures of catastrophe sets and their associated shock waves will lead to a deeper understanding of the geometries of biological forms in general.

The above example can be improved by incorporating a time-factor. Instead of V_C for each cell *C* we will take $V_{C,t}$ for the cell *C* at time *t.* Then the model has the same variables $(x_1, x_2, ..., x_n)$, but the control space becomes the region of 4-space (3 space, 1 time) corresponding to the evolution of the tissue through time. The catastrophe and shock wave sets *K,S* will now be subsets of 4-space. As the piece of tissue moves through time it will intersect *K* and *S* in various geometrical forms which will change with time. This is the basis for Thom's application of catastrophe theory to morphogenesis.

17.6.   The main theorem

Now we are ready to turn to Thom's theorem.

THEOREM.   For systems governed by smooth functions with at most four parameters **(but any number of variables)** *there are essentially only seven possible types of local geometric structure for stable catastrophe sets.*

Some of these words have to be explained.

Thom provides a list of seven explicit functions (*catastrophe models*) with the property that any other function $V_c(x)$ which has

1. at most four parameters, and
2. a catastrophe set *K* which is stable,

can, in a neighbourhood of any point *c* in *K,* be converted essentially (*see below*) into one of these seven catastrophe models by means of a smooth change of coordinates which also converts *K* essentially into the catastrophe set of the model. The change of coordinates may also involve adding a constant to V_c for each c, but of course this makes no difference to the critical points of V_c which are what concern us here. Finally, to obtain one of Thom's explicit functions it may be necessary to replace V_c by $-V_c$. Again this makes no difference to the configuration of critical points, but it converts maxima to minima and *vice versa* and so is important in applications.

It is frequently possible to include extra variables in each model without altering the catastrophe set. For example, we can do this by adding on a term of the form

$$\sum_{i=1}^{m} \pm y_i^2$$

where the y_i are extra variables and m can be as large as we like. To see that this makes no difference to K we simply observe that

$$\frac{\partial V_c}{\partial y_i} = 0$$

precisely when $y_i = 0$, and the enlarged matrix of second derivatives has zero determinant precisely when the original one does (all we have done is to augment it with diagonal elements, each of which is ± 2). Therefore the equation for the catastrophe set is not changed, and the whole catastrophe is taking place in the space of the original variables, with the new variables $y_1, y_2, ..., y_m$ all zero. For an illustration see Example (7) below.

It might happen that for a certain point c in K there are two or more distinct coalescences of critical points of V_c occurring at the same time, so that c really lies on the intersection of several catastrophe sets which are locally independent (although they may join together as part of a larger catastrophe set elsewhere in the control space). In this case Thom's list will not tell us what the local configuration looks like, but it will describe locally each of the component parts which pass through c.

The word 'essentially' in the theorem means that we must make all these allowances when trying to interpret a given catastrophe as one of Thom's seven.

17.7. Thom's list of seven catastrophe models

1. THE FOLD

$$V_{c_1}(x) \equiv x^3 + c_1 x$$

For $c_1 < 0$ there is one minimum and one maximum; for $c_1 > 0$

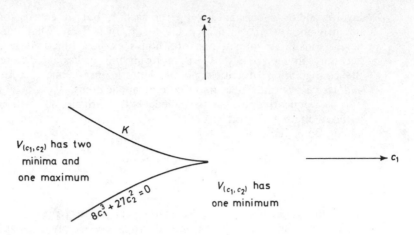

If we plot in (c_1, c_2, x)-space all the points such that x is a critical point of $V_{(c_1, c_2)}$ we obtain a surface S with a 'pleat':

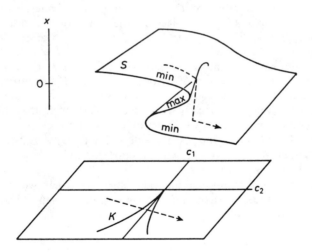

The points on S where the tangent plane is vertical are the points which project down to the cusp K in the (c_1, c_2)-plane. Using S we can see how equilibria switch as we vary the parameters (c_1, c_2), i.e. as we move in the plane.

Figure 17.4   The cusp catastrophe

there are no critical points. The catastrophe set K consists of the one point $c = 0$. (Compare the catastrophe in Example (5).)

If we plot critical points of V_{c_1} against c_1 we obtain the parabola $3x^2 + c_1 = 0$ in the (c_1, x)-plane. The branch of the parabola with $x > 0$ corresponds to the minimum of V_{c_1}, and as c_1 passes through 0 from below the minimum disappears and the equilibrium 'falls off' the curve towards $x = -\infty$. This indicates why this catastrophe is called the fold.

2. THE CUSP

$$V_{(c_1, c_2)}(x) \equiv x^4 + c_1 x^2 + c_2 x$$

Critical points occur where $V'_{(c_1, c_2)}(x) = 0$, i.e.

$$4x^3 + 2c_1 x + c_2 = 0$$

and they coalesce where $V''_{(c_1, c_2)}(x) = 0$, i.e.

$$12x^2 + 2c_1 = 0$$

Eliminating x gives

$$8c_1{}^3 + 27c_2{}^2 = 0$$

for the equation of the catastrophe set K, which is easily verified to be a cusp in the (c_1, c_2)-plane as in *Figure 17.4*. For $8c_1{}^3 + 27c_2{}^2 > 0$ there is just one critical point (a minimum) for $V_{(c_1, c_2)}$; for $8c_1{}^3 + 27c_2{}^2 < 0$ there are two minima and one maximum.

For the next three models the calculations become very involved, so we will simply give the functions and sketch the catastrophe sets K in the control space, indicating the nature of the critical points of V_c for c in each region complementary to K. The control space has dimension three in all cases.

3. THE SWALLOW-TAIL

$$V_{(c_1, c_2, c_3)}(x) \equiv x^5 + c_1 x^3 + c_2 x^2 + c_3 x$$

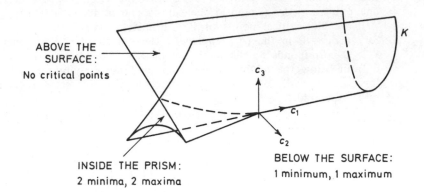

ABOVE THE SURFACE:
No critical points

INSIDE THE PRISM:
2 minima, 2 maxima

BELOW THE SURFACE:
1 minimum, 1 maximum

Figure 17.5   The swallow-tail

4.   THE ELLIPTIC UMBILIC, OR HAIR

$$V_{(c_1, c_2, c_3)}(x_1, x_2) \equiv x_1^3 - 3x_1 x_2^2 + c_1(x_1^2 + x_2^2) + c_2 x_1 + c_3 x_2$$

5.   THE HYPERBOLIC UMBILIC, OR BREAKING WAVE

$$V_{(c_1, c_2, c_3)}(x_1, x_2) \equiv x_1^3 + x_2^3 + c_1 x_1 x_2 + c_2 x_1 + c_3 x_2$$

The final two models each have control spaces of dimension four, so it is impossible to sketch the catastrophe sets. We simply give the functions $V_{(c_1, c_2, c_3, c_4)}$ in the two cases.

6.   THE BUTTERFLY

$$V_{(c_1, c_2, c_3, c_4)}(x) \equiv x^6 + c_1 x^4 + c_2 x^3 + c_3 x^2 + c_4 x$$

7.   THE PARABOLIC UMBILIC, OR THE MUSHROOM

$$V_{(c_1, c_2, c_3, c_4)}(x_1, x_2) \equiv x_2^4 + x_1^2 x_2 + c_1 x_1^2 + c_2 x_2^2 +$$

$$c_3 x_1 + c_4 x_2$$

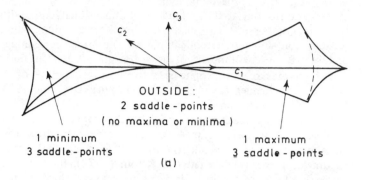

OUTSIDE:
2 saddle-points
(no maxima or minima)

1 minimum
3 saddle-points

1 maximum
3 saddle-points

(a)

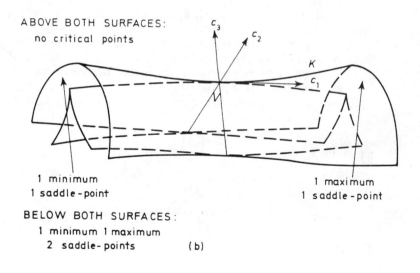

ABOVE BOTH SURFACES:
no critical points

1 minimum
1 saddle-point

1 maximum
1 saddle-point

BELOW BOTH SURFACES:
1 minimum 1 maximum
2 saddle-points

(b)

Figure 17.6 (a) The elliptic umbilic; (b) the hyperbolic umbilic

Some idea of the geometrical structure of the catastrophe sets for the butterfly and parabolic umbilic can be gained by taking sections of (c_1, c_2, c_3, c_4)-space by hyperplanes $c_1 = 0$, $c_2 = 0$ etc. The butterfly turns out to be a kind of 4-dimensional cusp with smaller cusps superimposed. The parabolic umbilic has some 3-dimensional sections containing hyperbolic umbilics and some containing elliptic umbilics: it can be thought of as a transition between the two.

For pictures of catastrophe sets from various points of view see Bröcker and Lander (1974), Callahan (1974), Godwin (1971), Thom (1972), Woodcock and Poston (1974).

Corresponding to any picture of a catastrophe set there is in theory an associated shock-wave picture. This is easy to draw for the cusp (it is just the line of symmetry $c_2 = 0$ with $c_1 < 0$) and quite easy to draw for the swallow-tail (*see* Thom (1972), p.67 of the English version). For the butterfly it is more involved, having an interesting geometric form which Thom relates to the formation of pockets (*ibid*, pp.69–73). The elliptic and hyperbolic umbilics have no shock waves since they do not produce more than one minimum. The shock wave set for the parabolic umbilic does not yet seem to have been given a good description.

Note: All the functions which appear in the list are in fact polynomials. However, it is important to realise that it is certainly *not* necessary for smooth functions to be polynomials in order for Thom's theorem to apply to them. It is precisely in this reduction of general smooth functions to polynomial form (under the assumption of stability for the catastrophe set K) that the power of the mathematics lies.

The next two examples illustrate some points arising from the theorem.

EXAMPLE 7

Let

$$V_{(c_1, c_2)}(x_1, x_2) \equiv x_1{}^3 - 2x_1 x_2 + c_1 x_1{}^2 + x_2{}^2 + c_2 x_1$$

The critical points are given by

$$\frac{\partial V_{(c_1, c_2)}}{\partial x_1} \equiv 3x_1{}^2 - 2x_2 + 2c_1 x_1 + c_2 = 0$$

$$\frac{\partial V_{(c_1, c_2)}}{\partial x_2} \equiv -2x_1 + 2x_2 = 0$$

i.e. $x_1 = x_2$ and $3x_1^2 + 2(c_1 - 1)x_1 + c_2 = 0$. Critical points coalesce where the determinant of second derivatives

$$\begin{vmatrix} 6x_1 + 2c_1 & -2 \\ -2 & 2 \end{vmatrix}$$

vanishes, i.e. $2(6x_1 + 2c_1) - 4 = 0$ or $x_1 = 1/3\,(1 - c_1)$. Substituting this in the above quadratic equation for x_1 we get

$$c_2 = \frac{1}{3}(c_1 - 1)^2$$

so the catastrophe set K is a parabola in the (c_1, c_2)-plane. In fact it is a stable catastrophe set, although we will not attempt to give a proof. Thom's theorem therefore implies that there is a smooth change of coordinates taking this model essentially to one of the seven. To find appropriate new coordinates in general is not an easy problem, particularly when the functions are not polynomials, since the theory invokes abstract results about the existence of solutions to systems of differential equations. Here we will simply claim (and leave the verification as an exercise) that the new coordinates (u_1, u_2), (γ_1, γ_2) given by

$$u_1 = x_1 + \frac{1}{3}(c_1 - 1)$$

$$u_2 = x_1 - x_2$$

$$\gamma_1 = c_2 - \frac{1}{3}(c_1 - 1)^2$$

$$\gamma_2 = c_1 - 1$$

convert $V_{(c_1, c_2)}(x_1, x_2)$ into

$$W_{(\gamma_1, \gamma_2)}(u_1, u_2) \equiv u_1^3 + \gamma_1 u_1 + u_2^2 - \frac{1}{3}\gamma_2\left(\gamma_1 + \frac{1}{9}\gamma_2^2\right)$$

In rescaling the $W_{(\gamma_1, \gamma_2)}$-axis (for each (γ_1, γ_2)) by adding on

$$\frac{1}{3}\gamma_2\left(\gamma_1 + \frac{1}{9}\gamma_2^2\right)$$

this becomes

$$\hat{W}_{(\gamma_1, \gamma_2)}(u_1, u_2) \equiv u_1^3 + \gamma_1 u_1 + u_2^2$$

which is essentially (i.e. allowing for the unimportant $u_2{}^2$ term and the irrelevance of γ_2) the model for the fold catastrophe. Observe that the change of coordinates has converted the catastrophe set K in the (c_1, c_2)-plane into the straight line $\gamma_1 = 0$ in the (γ_1, γ_2)-plane.

EXAMPLE 8

Let

$$V_{(c_1, c_2)}(x) \equiv x_1{}^3 + (c_1{}^2 + c_2{}^2)\, x_1$$

The catastrophe set K is easily seen to consist of the single point $(0,0)$ in the (c_1, c_2)-plane: for all other points $c_1{}^2 + c_2{}^2 > 0$, and so $V_{(c_1, c_2)}$ has no critical points. This is *not* essentially either of the two catastrophes with two parameters which occur in Thom's list (the cusp, or the fold with one of the parameters playing no rôle). The only explanation (according to Thom's theorem) is that K cannot be stable, i.e. by changing V and all its derivatives by an arbitrarily small amount we must be able to change the structure of K. It is easy to see that we can indeed do this by replacing $V_{(c_1, c_2)}$ by $V^\epsilon{}_{(c_1, c_2)}$ given by

$$V^\epsilon{}_{(c_1, c_2)}(x) \equiv x_1{}^3 + (c_1{}^2 + c_2{}^2 - \epsilon)\, x_1$$

for any ϵ. When $\epsilon < 0$ the catastrophe set is empty; when $\epsilon > 0$ it is the circle $c_1{}^2 + c_2{}^2 = \epsilon$. In both cases the catastrophe set can be shown to be stable.

REMARKS

1. The catastrophes occurring in the Zeeman catastrophe machine should, according to this theory, be locally describable by one of Thom's models with two parameters. It is worthwhile verifying experimentally (provided the machine is in efficient working order) that the catastrophe set K is a curve in the control space with four cusps as shown by Poston and Woodcock (1973).
2. Smooth changes of coordinates can bend straight lines into curves, but cannot make two lines tangent which were not tangent before, and cannot introduce corners. This means that the three curves in *Figure 17.7* cannot be transformed into each other by smooth changes of coordinates. Hence, for example, the cusp in model 2 is essentially different from the parabola in Example (5).

Figure 17.7

17.8. Use of the catastrophe models in applications

Catastrophe theory or, more accurately, the *catastrophe method,* provides, as we have seen, a way of modelling some of the discontinuities which can occur in evolving processes of many different kinds. It fits various types of discontinuity into a coherent mathematical framework and, if it does nothing more, it certainly provides a satisfying theoretical unification of some apparently diverse phenomena. Moreover, it gives a vivid insight into the *qualitative* nature of these discontinuities, showing that there are certain clearly-recognisable geometric forms (cusp, swallow-tail etc.) which control the discontinuous behaviour of systems governed by smooth functions with $k \leqslant 4$ parameters but with *any number of variables.*

Nevertheless, an applied mathematician is entitled to ask whether catastrophe theory can tell us anything about real life that we did not know before in some other guise. It is all very well being able to appreciate the beautiful underlying geometry of a situation, but does this appreciation allow us to make predictions of behaviour that we were not able to make previously? We will consider now some possible answers to these challenges.

1. Qualitative insight and understanding may often be more valuable than technical quantitative prediction (*see* Introduction), and in any case make an essential launching-pad for further advances. It is necessary to have a global picture of what is happening in order to progress significantly in understanding specific aspects. This is the central theme throughout much of Thom's writing, with particular relevance to models of biological development (Thom, 1972, 1973).

2. If discontinuities in the behaviour of a given dynamical system exhibit certain geometrical features that suggest the presence of one of Thom's catastrophes, then it is reasonable to make the *conjecture* that the system is in some way governed by a smooth function with at most four parameters. The theory then tells us that up to suitable change of variables and parameters the smooth function must be essentially (with the technical meaning of Section 17.6) one of the

seven specified in Thom's list of catastrophes. We therefore encourage the physicist, biologist, sociologist etc. to look for one or two key variables x or (x_1, x_2) and the $k \leqslant 4$ key parameters which govern that particular catastrophe, and assure him that the rest of his data will be irrelevant to the qualitative study and prediction of the discontinuities in the system. This is exploited very much by Zeeman in his appealing applications of catastrophe theory to sociological and psychological phenomena (Isnard and Zeeman, 1975; Zeeman, 1971, 1974. *See also* Zeeman, 1976).

3. Suppose we are studying a phenomenon which we already know to be modelled (after, perhaps, a change of variables) by a particular catastrophe. For example, we might be looking at the coalescence of the two minima of $x^4 + \lambda x^2 + \mu x$ as we vary λ and μ. We could then use known geometry of the catastrophe (in this case the *cusp* and its associated surface in 3-space; *see Figure 17.4*) to obtain more precise qualitative information about the way the minima of the function vary as we move the parameters about in the control space. Of course, such information could also be obtained, if we knew what to look for, by doing calculations with the formula for the function, but it would not leap to the eye as it would from the geometrical model. It is this aspect of catastrophe modelling that so far seems to be most useful in structural engineering (Sewell, 1976; Thompson, 1975).

4. An important feature of the theory is that it deals with catastrophe sets which are themselves stable, i.e. those which are not perturbed by arbitrarily small perturbations of the whole system. This means that if we constructed a catastrophe model for a real-life system known to be governed by a family of smooth functions, and we found that the catastrophe set K in the control space was *not* essentially one of Thom's seven types, though the number of controls was nevertheless $\leqslant 4$, then we could say that the model itself must be unstable, and therefore not the right one for modelling a real system. We could make positive suggestions for improving the model by including a small perturbation term which would make it stable, as in Example (8). These ideas have been applied usefully by Berry (1976) in work on refraction of light, and by Chillingworth and Furness (1975) in studying a simple dynamo model for the behaviour of the earth's magnetic field.

17.9. Conclusion

As the ideas of catastrophe theory develop, more and more mathematical models of real-life phenomena involving catastrophes are continually being suggested. The subject is still in its infancy, and although it remains to be seen to what extent Thom's work will produce real advances in the understanding of natural processes, his insights have already profoundly influenced the thinking of mathematicians interested in modelling dynamical systems. For the present an important task is to assemble more examples of systems with discontinuities modelled by the known catastrophes (Thom's list can in fact now be extended to classify many catastrophes with $k > 4$ parameters). To do this we must contemplate the geometric forms involved, and learn to recognise them in their manifestations in the world around us. A knowledge of these forms helps us to interpret many everyday phenomena, such as a mushroom, or a loss of temper, or the cusp-like *caustic* of reflected light on the surface of a cup of tea, in a new and unexpected way.

17.10. References

BERRY, M.V. (1976). 'Waves and Thom's Theorem', *Advances in Physics*, **25**, 1

BRÖCKER, Th. and LANDER, L. (1974). *Differentiable Germs and Catastrophes*, London Mathematical Society Lecture Notes, **17**,

CALLAHAN, J.J. (1974). 'Singularities and Plane Maps', *Amer. Math. Monthly*, **81**, 211

CHILLINGWORTH, D.R.J. and FURNESS, P.M.D. (1975). *Dynamical Systems – Warwick 1974*. Ed. Manning, A. Springer Lecture Notes in Mathematics, **468**, 91

GODWIN, A.N. (1971). 'Three-dimensional Pictures for Thom's Parabolic Umbilic', *Publ. Math. IHES*, **40**, 117

ISNARD, C.A. and ZEEMAN, E.C. (1975). 'Some Models from Catastrophe Theory in the Social Sciences', in *Use of Models in the Social Sciences*. Ed. Collins, L. Tavistock; London

PAVLIDIS, Th. (1973). *Biological Oscillators: Their Mathematical Analysis*. Academic Press; New York

POSTON, T. and WOODCOCK, A.E.R. (1973). 'Zeeman's Catastrophe Machine', *Proc. Cambridge Philos. Soc.*, **74**, 217

ROSEN, R. (Ed.) (1972). *Foundations of Mathematical Biology, II*. Academic Press; New York

ROSEN, R. (1970). *Dynamical System Theory in Biology*. Wiley; New York

SEWELL, M.J. (1976). 'Some Mechanical Examples of Catastrophe Theory', *Bull. Inst. Math. Appl.*, **12**, 163

THOM, R. (1972). *Stabilité Structurelle et Morphogénèse*. Benjamin; Reading, Massachusetts

THOM, R. (1973). 'A Global Dynamical Scheme for Vertebrate Embryology', *Some Mathematical Questions in Biology VI*, *Amer. Math. Soc.*, 3

THOM, R. (1974). *Modèles Mathématiques de la Morphogénèse*, Union Générale d'Editions ('10/18' series); Paris

THOMPSON, J.M.T. (1975). 'Experiments in Catastrophe', *Nature*, **254**, 392

THOMPSON, J.M.T. and HUNT, G.W. (1973). *A General Theory of Elastic Stability*. Wiley; New York

TIMOSHENKO, S. (1936).   *Theory of Elastic Stability.*   McGraw-Hill; New York

WOODCOCK, A.E.R. and POSTON, T. (1974).   *A Geometrical Study of the Elementary Catastrophes,* Springer Lecture Notes in Mathematics, **373**

ZEEMAN, E.C. (1971).   'The Geometry of Catastrophe', *The Times Literary Supplement,* 1556

ZEEMAN, E.C. (1972).   'A Catastrophe Machine'.   In *Towards a Theoretical Biology Vol.IV.*   Ed. Waddington, C.H.   Edinburgh University Press

ZEEMAN, E.C. (1974).   'On the Unstable Behaviour of Stock Exchanges', *J. Math. Economics,* · **1**, 39

ZEEMAN, E.C. (1976).   'Catastrophe Theory', *Scientific American,* **234**, 65

FURTHER READING

There is to date no better overall reference for applications of catastrophe theory than the excellent survey by Zeeman (1976).   However, to gain an idea of depth and subtlety of the underlying philosophy, as well as the potentially vast scope of catastrophe modelling in the study of natural phenomena, it is essential at least to browse through Thom's book (Thom, 1972).   The interview with Thom in *The Times Higher Educational Supplement (5.xii.75)* is also worth reading.

Other articles of a general nature include:

CHILLINGWORTH, D.R.J. (1975).   'Elementary Catastrophe Theory', *Bull. Inst. Math. Appl.,* **11**, 155

STEWART, I.N. (1975).   'The Seven Elementary Catastrophes', *New Scientist,* **68**, 447

SUSSMAN, H.J. (1975).   'Catastrophe Theory', *Synthèse,* **31**, 229

THOM, R. and ZEEMAN, E.C. (1975).   'Catastrophe Theory: Its Present State and Future Perspectives', *Dynamical Systems – Warwick 1974.* Ed. Manning, A.   Springer Lecture Notes in Mathematics, **468**, 366

17.11.   Problems for further study

1.   The equation of motion of a pendulum consisting of a particle of mass m on the end of a weightless rod of length l is

$$\ddot{\theta} + \frac{b}{ml}\,\dot{\theta} + \frac{g}{l}\,\sin\theta \;=\; 0 \qquad\qquad (*)$$

where θ = angular displacement of rod from vertical, $\dot{\theta} = d\theta/dt$, and b is a positive constant (representing frictional retardation).   Writing $x_1 = \theta$, $x_2 = \dot{\theta}$, express (*) as a pair of first-order differential equations and find a function $V(x_1, x_2)$ which governs this dynamical system.   (Hint: consider the total energy of the system.)

2. Regarding Q and m as control parameters in Example 3, find the catastrophe set K. Does such a catastrophe appear in Thom's list? If so, which one is it? If not, why not?

3. Consider a system governed by the smooth function

$$V_{(c_1, c_2)}(x) \equiv x^4 + c_1 x^2 + c_2 x$$

Suppose the point $c = (c_1, c_2)$ in the control space is made to oscillate with time t according to the equation

$$c(t) = (-1, A \cos 2\pi t)$$

How will the observed (equilibrium) value of x vary (a) when A is small, (b) when A is large, assuming (in each case) (i) the delay situation, (ii) the Maxwell situation?

4. Let K be the swallow-tail catastrophe set (*Figure 17.6*). At some points K looks locally like the intersection of two planes meeting approximately at right angles, although this configuration does not occur as one of those on Thom's list. What is the explanation for this apparent contradiction?

5. Suppose you are studying a dynamical system governed by a smooth function with two parameters (c_1, c_2), and you find that the catastrophe set consists of the two semi-axes $c_1 = 0$, $c_2 > 0$ and $c_2 = 0$, $c_1 > 0$. What conclusion can you draw about the suitability of this system as a model for a real-life system? Consider various 2-dimensional sections of the hyperbolic umbilic catastrophe set, and suggest how the above model might be improved.

6. A constant function $V(x) \equiv b$ is not stable, in that by arbitrarily small changes in V and V' it is possible to convert V into a function having any number of maxima and minima. However, constant functions are obviously useful in practice. This seems to be inconsistent with the hypothesis of structural stability. Discuss the relevance of structural stability, with this example in mind.

7. Explore the mathematical ideas that underlie the use of the word *stable* in each of the following: *stable equilibrium, stable catastrophe, stable economy, stable personality.*

8. The Zeeman Catastrophe Machine is really a large-size mechanical *switch*. (The word *switch* itself means a device which converts a smooth input into a discontinuous output). Explore how more elaborate switching devices might be constructed using catastrophes other than the cusp. Give a catastrophe-theoretical description of a biological switch, in which the gradual increase in concentration of one chemical agent can cause the sudden onset of a certain cellular reaction.

9.   A system governed by a smooth function V cannot show any *periodic* behaviour, since (except at equilibrium points) $V(x(t))$ is always decreasing as t increases.   In general, though, dynamical systems may have periodic evolutions from suitable starting-points.   Think how the idea of *equilibrium point* and *stable equilibrium* might be generalised, and then try to formulate a definition of *catastrophe* in this wider context.

10.   Discuss the importance of *qualitative* versus *quantitative* results in applied mathematics, bearing in mind the remarks in Section 17.8. How far do you accept the outline given here of Thom's point of view?

INDEX

Ablation, 73

Bernoulli's equation, 39

Calculus of Variations, 17
Catastrophe theory, 231
Conservation of energy, 73, 74
Conservation of mass, 40
Conservation of momentum, 30
Control models, 12, 231

Diabetes mellitus, differential
  model of, 116
Dynamic programming, 151, 154

Eigenvalues, 68
Extremum principle, 92

Finite elements, 84
First order ordinary differential
  equations, 19, 30, 43, 73, 78,
  79, 99, 100, 102, 104, 107,
  108, 117, 118, 189, 234
Free boundary problems, 46, 73

Game theory, 183
Grade structure, control of, 161
Graph theory, 58

Helmholtz-Kirchoff free streamline theory, 39

Laplace transforms, 78, 132
Laser drilling, 71

Least-squares fit, 65
Life-tables, 103
Linear programming, 188

Markov chains, 165
Matrix models, 57, 87, 165, 219
Mine warfare, 183
Minimal paths, 20
Molecular models, 56
Motor insurance, model for, 174

Network flow models, 199
Newton's laws of motion, 19, 28

Optimisation, 155, 171, 208

Partial differential equations, 74
Perturbation theory, 76
Planning models, 143, 161, 175
Poisson distribution, 130, 175
Predator-prey models, 104
Probability distributions, 130,
  175, 186

Road traffic models, 127

Stability, 231
Steering models, 12
Stefan problems, 74
Stochastic models, 127, 193
Stocks and flows, 161, 175, 211
Stress analysis, 83

Taylor's theorem, 30

259

Three-stage rockets, 26
Transportation models, 211

University salaries, 161
Urban structure, model of, 217

Validation of models, 124, 140
Variational calculus, 17

Zeeman's catastrophe machine, 239
Zero-sum games, 184